Progress in Mathematics
Volume 41

Richard P. Stanley

Combinatorics
and
Commutative Algebra

Second Edition

Birkhäuser

Boston • Basel • Berlin

Richard P. Stanley
Department of Mathematics
Massachusetts Institute of Technology
Cambridge, MA 02139

Library of Congress Cataloging-in-Publication Data

Stanley, Richard P., 1944-
    Combinatorics and commutative algebra / Richard P. Stanley. -- 2nd
ed.
        p.   cm. -- (Progress in mathematics ; v. 41)
        Includes bibliographical references.
        ISBN 0-8176-3836-9 (alk. paper). -- ISBN 3-7643-3836-9 (alk.
paper)
            1. Commutative algebra. 2. Combinatorial analysis.   I. Title.
    II. Series: Progress in mathematics (Boston, Mass.) ; vol. 41
        QA251.3.S72   1996   95-25196
        512'.24--dc20   CIP

Printed on acid-free paper
© 1996 Birkhäuser Boston, 2nd ed.   *Birkhäuser*
    1983 Birkhäuser Boston, 1st ed.

ISBN 0-8176-3836-9
ISBN 3-7643-3836-9
Layout, typesetting by TeXniques, Boston, MA
Printed and bound by Quinn-Woodbine, Woodbine, NJ
Printed in the United States of America

9 8 7 6 5 4 3 2 1

Contents

Preface to the Second Edition

Since the appearance of the first edition many further developments have taken place in the area of "combinatorial commutative algebra." Perhaps the most interesting advances concern the face ring of a simplicial complex, the subject of Chapter 2. Therefore I have added an additional chapter summarizing new work in this area. It provides strong additional evidence of the felicitous symbiosis between the subjects of combinatorics and commutative algebra. I have also added a collection of exercises taken from a course taught at M.I.T. Chapters 0-2 have been corrected and brought up-to-date in only minor ways.

I am grateful to the staff at Birkhäuser for their help in preparing this new edition. Ann Kostant in particular has been an ideal editor, while Sarah Jaffe has done an excellent job of TEXing the original text of the first edition and merging the list of references there with the many new references. Finally I wish to thank the numerous persons who have contributed valuable suggestions concerning the material in Chapter 3, including Ron Adin, Louis Billera, Anders Björner, Art Duval, David Eisenbud, Takayuki Hibi, Tony Iarrobino, Gil Kalai, and Christian Peskine.

Richard Stanley
Cambridge, Massachusetts
October 20, 1995

Preface to the First Edition

These notes are based on a series of eight lectures given at the University of Stockholm during April and May, 1981. They were intended to give an overview of two topics from "combinatorial commutative algebra," viz., (1) solutions to linear equations in nonnegative integers (which is equivalent to the theory of invariants of a torus acting linearly on a polynomial ring), and (2) the face ring of a simplicial complex. In order to give a broad perspective many details and specialized topics have been regretfully omitted. In general, proofs have been provided only for those results which were obscure or inaccessible in the literature at the time of the lectures. The original lectures presupposed considerable background in commutative algebra, homological algebra, and algebraic topology. In order to broaden the accessibility of these notes, Chapter 0 has been prepared with the kind assistance of Karen Collins. This chapter provides most of the background information in algebra, combinatorics, and topology needed to read the subsequent chapters.

I wish to express my gratitude to the University of Stockholm, in particular to Jan-Erik Roos, for the kind invitation to visit in conjunction with the year devoted to algebraic geometry and commutative algebra at the Institut Mittag-Leffler. I am also grateful for the many insightful comments and suggestions made by persons attending the lectures, including Anders Björner, Ralf Fröberg, Christer Lech, and Jan-Erik Roos. Special appreciation goes to Anders Björner for the time-consuming and relatively thankless task of writing up these lecture notes. Finally I wish to thank Maura A. McNiff and Ruby Aguirre for their excellent preparation of the manuscript.

Richard Stanley
Cambridge, Massachusetts
May, 1983

Notation

$\mathbb{C}$	complex numbers
$\mathbb{N}$	nonnegative integers
$\mathbb{P}$	positive integers
$\mathbb{Q}$	rational numbers
$\mathbb{R}$	real numbers
$\mathbb{Z}$	integers
$\mathbb{R}^+$	nonnegative real numbers
$[n]$	for $n \in \mathbb{N}$, the set $\{1, 2, \ldots, n\}$
N-matrix	a matrix whose entries belong to the set N
$N[x]$	polynomials in x whose coefficients belong to the set N
$N[[x]]$	formal power series in x whose coefficients belong to the set N
$\#S$	cardinality of the finite set S
$\|\cdot\|$	cardinality or geometric realization, according to context
$T \subseteq S$	T is a subset of S
$T \subset S$	T is a subset of S and $T \neq S$
$\alpha > 0$	for a vector $\alpha = (\alpha_1, \ldots, \alpha_n) \in \mathbb{R}^n$, this means $\alpha_i > 0$ for all i
k^*	nonzero elements of the field k
kE	vector space over k with basis E
$\cong$	symbol for isomorphism
$\approx$	symbol for homeomorphism
$\oplus, \coprod$	direct sum (of vector spaces or modules)
im f	image $f(M)$ of the homomorphism $f : M \to N$
ker f	kernel of $f : M \to N$
vol $\mathcal{P}$	volume (= Lebesgue measure) of the set $\mathcal{P} \subseteq \mathbb{R}^n$
δ_{ij}	the Kronecker delta (= 1 if $i = j$, and = 0 if $i \neq j$)

Chapter 0

Background

1 Combinatorics

The purpose of this introduction is to provide the reader with the relevant background from combinatorics, algebra, and topology for understanding of the text. In general the reader may prefer to begin with Chapter 1 and refer back to this chapter only when necessary. We assume the reader is familiar with standard (first-year graduate) material but has no specialized knowledge of combinatorics, commutative algebra, homological algebra, or algebraic topology.

We begin with a discussion of rational power series in one variable [162, Ch. 4]. Let $F(x) = \sum_{n \geq 0} f(n)x^n \in \mathbb{C}[[x]]$ be a formal power series with complex coefficients. We say $F(x)$ is *rational* if there exist polynomials $P(x), Q(x) \in \mathbb{C}[x]$ for which $F(x) = P(x)/Q(x)$, i.e., $F(x)Q(x) = P(x)$ in the ring $\mathbb{C}[[x]]$. Without loss of generality we may assume $Q(0) = 1$. Define $\deg F(x) = \deg P(x) - \deg Q(x)$.

1.1 Theorem. *Let* $\alpha_1, \alpha_2, \ldots, \alpha_d$ *be a fixed sequence of complex numbers, $d \geq 1$ and $\alpha_d \neq 0$. The following conditions on a function $f : \mathbb{N} \to \mathbb{C}$ are equivalent:*

(i)
$$\sum_{n \geq 0} f(n)x^n = \frac{P(x)}{Q(x)} ,$$

where $Q(x) = 1 + \alpha_1 x + \cdots + \alpha_d x^d$ and $P(x)$ is a polynomial in x of degree less than d.

(ii) *For all $n \geq 0$,*

$$f(n+d) + \alpha_1 f(n+d-1) + \cdots + \alpha_d f(n) = 0 . \tag{1}$$

(iii) *For all $n \geq 0$,*

$$f(n) = \sum_{i=1}^{k} P_i(n)\gamma_i^n ,\qquad(2)$$

where $1 + \alpha_1 x + \cdots + \alpha_d x^d = \prod_{i=1}^{k}(1 - \gamma_i x)^{d_i}$, the γ_i's are distinct, and $P_i(n)$ is a polynomial in n of degree less than d_i.

Sketch of proof. Fix $Q(x) = 1 + \alpha_1 x + \cdots + \alpha_d x^d$. Define four complex vector spaces as follows:

$$V_1 = \{f : \mathbb{N} \to \mathbb{C} \ \text{ such that (i) holds}\}$$
$$V_2 = \{f : \mathbb{N} \to \mathbb{C} \ \text{ such that (ii) holds}\}$$
$$V_3 = \{f : \mathbb{N} \to \mathbb{C} \ \text{ such that (iii) holds}\}$$
$$V_4 = \{f : \mathbb{N} \to \mathbb{C} \ \text{ such that } \sum_{n\geq0} f(n)x^n = \sum_{i=1}^{k} G_i(x)(1 - \gamma_i x)^{-d_i} ,$$

for some polynomials $G_i(x)$ of degree less than d_i, where γ_i and d_i have the same meaning as in (iii)$\}$

It is easily seen that $\dim V_j = d$ for $1 \leq j \leq 4$. One readily shows $V_1 \subseteq V_2$, $V_4 \subseteq V_1$, $V_4 \subseteq V_3$. Hence $V_1 = V_2 = V_3(= V_4)$. □

We next consider rational functions $F(x) = P(x)/Q(x)$ with $\deg P \geq \deg Q$, i.e., $\deg F(x) \geq 0$.

1.2 Proposition. *Let $f : \mathbb{N} \to \mathbb{C}$ and suppose that*

$$\sum_{n\geq0} f(n)x^n = P(x)/Q(x)$$

where $P, Q \in \mathbb{C}[x]$. Then there is a unique finite set $E_f \subset \mathbb{N}$ (the exceptional set *of f) and a unique function $f_1 : E_f \to \mathbb{C}^* = \mathbb{C} - \{0\}$ such that the function $g : \mathbb{N} \to \mathbb{C}$ defined by*

$$g(n) = \begin{cases} f(n) , & \text{if } n \notin E_f \\ f(n) + f_1(n) , & \text{if } n \in E_f , \end{cases}$$

satisfies $\sum_{n\geq0} g(n)x^n = R(x)/Q(x)$, where $R \in \mathbb{C}[x]$ and $\deg R < \deg Q$. Moreover, assuming $E_f \neq \emptyset$ (i.e., $\deg P \geq \deg Q$), define $m(f) = \max\{i : i \in E_f\}$. Then:

(i) *$m(f) = \deg P - \deg Q$,*

(ii) *$m(f)$ is the largest integer n for which (1) fails to hold,*

(iii) *Writing $Q(x) = \prod_1^k (1-\gamma_i x)^{d_i}$ as above, there are unique polynomials $P_1, \ldots, P_k$ for which (2) holds for n sufficiently large. Then $m(f)$ is the largest integer n for which (2) fails.*

Sketch of proof. By the division algorithm for $\mathbb{C}[x]$, there are unique polynomials $L(x)$ and $R(x)$ with $\deg R < \deg Q$, such that

$$\frac{P(x)}{Q(x)} = L(x) + \frac{R(x)}{Q(x)} \ .$$

We then define E_f, $g(n)$, and $f_1(n)$ by

$$\sum_{n \geq 0} g(n)x^n = \frac{R(x)}{Q(x)}$$

$$E_f = \{i : \text{the coefficient of } x^i \text{ in } L(x) \text{ is nonzero}\}$$

$$\sum_{n \in E_f} f_1(n)x^n = L(x) \ .$$

The proof follows easily. □

If $n \in E_f$, then we call $f(n)$ an *exceptional value* of f. Thus f has no exceptional values if and only if $\deg P(x)/Q(x) < 0$.

Two special cases of Theorem 1.1 will be of interest to us here.

1.3 Corollary. *Let* $f : \mathbb{N} \to \mathbb{C}$, *and let* $d \in \mathbb{N}$. *The following three conditions are equivalent:*

(i)
$$\sum_{n \geq 0} f(n)x^n = \frac{P(x)}{(1-x)^{d+1}} \ ,$$

where $P(x) \in \mathbb{C}[x]$ *and* $\deg P \leq d$.

(ii) *For all* $n \geq 0$,

$$\sum_{i=0}^{d+1} (-1)^i \binom{d+1}{i} f(n+i) = 0 \ .$$

(In the calculus of finite differences this condition is written $\Delta^{d+1}f(n) = 0$.*)*

(iii) $f(n)$ *is a polynomial function of* n *of degree at most* d. *(Moreover,* $f(n)$ *has degree exactly* d *if and only if* $P(1) \neq 0$. *In this case, the leading coefficient of* $f(n)$ *is* $P(1)/d!$.*)*

1.4 Definition. A *quasipolynomial* (known by many other names, such as "pseudo-polynomial" and "polynomial on residue classes") *of degree d* is a function $f : \mathbb{N} \to \mathbb{C}$ (or $f : \mathbb{Z} \to \mathbb{C}$) of the form

$$f(n) = c_d(n)n^d + c_{d-1}(n)n^{d-1} + \cdots + c_0(n) ,$$

where each $c_i(n)$ is a periodic function (with integer period), and where $c_d(n)$ is not identically zero. Equivalently, f is a quasipolynomial if there exists an integer $N > 0$ (viz., a common period of $c_0, c_1, \ldots, c_d$) and polynomials $f_0, f_1, \ldots, f_{N-1}$ such that

$$f(n) = f_i(n) \quad \text{if} \quad n \equiv i \ (\mathrm{mod}\, N) .$$

The integer N (which is not unique) will be called a *quasiperiod* of f.

1.5 Corollary. *The following conditions on a function $f : \mathbb{N} \to \mathbb{C}$ and integer $N > 0$ are equivalent:*

(i) *f is a quasipolynomial of quasiperiod N.*

(ii) *$\sum_{n \geq 0} f(n)x^n = \frac{P(x)}{Q(x)}$, where $P(x), Q(x) \in \mathbb{C}[x]$, every zero α of $Q(x)$ satisfies $\alpha^N = 1$ (provided $P(x)/Q(x)$ has been reduced to lowest terms), and $\deg P < \deg Q$.*

(iii) *For all $n \geq 0$,*
$$f(n) = \sum_{i=1}^k P_i(n)\gamma_i^n ,$$

where each P_i is a polynomial function of n and each γ_i satisfies $\gamma_i^N = 1$.

For instance (see Ch. 1, Cor. 4.2), if

$$\sum_{n \geq 0} f(n)x^n = \frac{V(x)}{(1 - x)^a(1 - x^2)^b} ,$$

where $V(x) \in \mathbb{C}[x]$ and $\deg V < a + 2b$, then $f(n)$ is a quasipolynomial of quasiperiod two. Thus there are polynomials $g(n), h(n)$ for which

$$f(n) = g(n) + (-1)^n h(n) .$$

We next discuss the problem of solving linear homogenous equations in nonnegative integers. Let Φ be an $r \times n$ $\mathbb{Z}$-matrix, and define

$$E_\Phi = \{\beta \in \mathbb{N}^n \mid \Phi\beta = 0\} .$$

Thus E_Φ is the set of solutions in nonnegative integers to the system $\Phi\beta = 0$ of homogeneous linear equations with integer coefficients. Clearly E_Φ forms a *submonoid* of $\mathbb{N}^n$, i.e., is closed under addition and contains 0. Chapter I is devoted primarily to the study of the monoid E_Φ and the related set $E_{\Phi,\alpha}$ of solutions to $\Phi\beta = \alpha$.

Hilbert deduced from his Basis Theorem (see Theorem 2.2 below) that E_Φ is a finitely-generated monoid. A simple direct proof was given by Gordan in 1900. If we define a formal power series

$$E_\Phi(x) = \sum_{\beta \in E_\Phi} x^\beta ,$$

where $x^\beta = x_1^{\beta_1} \cdots x_n^{\beta_n}$ and $\beta = (\beta_1, \ldots, \beta_n)$, then it was shown by Hilbert that $E_\Phi(x)$ is a rational function of x. In 1903 E. B. Elliott described an algorithm for computing $E_\Phi(x)$ and discussed some special cases. This algorithm was subsequently used by MacMahon [119, Sections VIII–X] to investigate a wide variety of combinatorial problems. In particular, in [119, Section VIII, Ch. VII] he computes the number $H_3(r)$ of 3×3 $\mathbb{N}$-matrices whose row and column sums are equal to r. (Such matrices are called *integer stochastic matrices* or *magic squares*.) Anand, Dumir, and Gupta [5] rediscovered MacMahon's result and made some conjectures (see Ch. I, Conj. 1.1(i)–(iii)) about the number $H_n(r)$ of $n \times n$ $\mathbb{N}$-matrices with line sum r. Stanley [147] proved the conjectures of Anand-Dumir-Gupta and generalized them to "magic labelings of graphs." This included a related conjecture of Carlitz on *symmetric* matrices with equal line sums. Stanley's proof was based upon the Elliott-MacMahon algorithm and the Hilbert syzygy theorem of commutative algebra (see Ch. I.11). In these notes we will prove these results using different tools from commutative algebra. An independent geometric proof was also given by E. Ehrhart [71].

Next we review some geometric notions. Some standard references are [86], [127], [44], and [188].

A (*convex*) *polytope* $\mathcal{P}$ is the convex hull of finitely many points in $\mathbb{R}^n$. Equivalently, $\mathcal{P}$ is a bounded intersection of finitely many half-spaces. A convex polytope is homeomorphic to a closed d-dimensional ball for some d, called the *dimension* of $\mathcal{P}$ and denoted $\dim \mathcal{P}$. A *supporting hyperplane* of $\mathcal{P}$ is an (affine) hyperplane $\mathcal{H}$ which intersects $\mathcal{P}$ and for which $\mathcal{P}$ is contained in one of the two closed half-spaces determined by $\mathcal{H}$. The intersection $\mathcal{P} \cap \mathcal{H}$ is a *face* of $\mathcal{P}$, and we also call the empty set $\emptyset$ a face (of dimension -1). A 0-dimensional face is a *vertex* of $\mathcal{P}$, and a $(d-1)$-dimensional face is a *facet*. Every face F of $\mathcal{P}$ is itself a convex polytope, and a face G of F is a face of $\mathcal{P}$. The *boundary* $\partial\mathcal{P}$ of $\mathcal{P}$ is the union of all faces of $\mathcal{P}$ except $\mathcal{P}$ itself. It coincides with the notion of boundary in the usual topological sense.

A polytope $\mathcal{P}^*$ is *dual* to $\mathcal{P}$ if there is an inclusion-reversing one-to-one correspondence $F \to F^*$ between the faces F of $\mathcal{P}$ and F^* of $\mathcal{P}^*$. Every polytope $\mathcal{P}$ has a dual, which we won't bother to construct here.

A *polyhedral complex* is a collection Γ of polytopes in $\mathbb{R}^n$ satisfying

(a) If $\mathcal{P} \in \Gamma$ and F is a face of $\mathcal{P}$, then $F \in \Gamma$, and

(b) If $\mathcal{P}, \mathcal{P}' \in \Gamma$, then $\mathcal{P} \cap \mathcal{P}'$ is a common face of $\mathcal{P}$ and $\mathcal{P}'$.

A (*convex*) *cone* $\mathcal{C}$ is a subset of $\mathbb{R}^n$ which is closed under the operation of taking *nonnegative* linear combinations. A cone $\mathcal{C}$ is *polyhedral* if it is the intersection of finitely many closed half-spaces (which must all contain 0 in their boundary). The *dimension* $\dim \mathcal{C}$ of a polyhedral cone $\mathcal{C}$ may be defined as the dimension of its linear span. We define a *face* of $\mathcal{C}$ as we did for polytopes, except that $\emptyset$ is never a face. If $\{0\}$ is a face, we call it the *vertex* of $\mathcal{C}$. A 1-dimensional face is an *extreme ray*.

2   Commutative algebra and homological algebra

Our basic references are [6], [92], [93], [120], and [121]. Hilton and Wu [93] is recommended in particular as a relatively painless introduction to homological algebra.

All our rings A are commutative with identity 1. In particular, every subring of A contains 1.

2.1 Definition. A ring A is said to be *noetherian* if it satisfies the following equivalent conditions:

(a) Every nonempty set of ideals in A has a maximal element.

(b) Every ascending chain $I_0 \subseteq I_1 \subseteq \cdots$ of ideals of A eventually stabilizes (i.e., $I_i = I_{i+1}$ for all large i).

(c) Every ideal of A is finitely-generated.

2.2 Hilbert Basis Theorem. *If A is noetherian, then the polynomial ring $A[x]$ is noetherian.*

2.3 Example. Clearly any field k is noetherian. By repeated application of the theorem, $k[x_1, \ldots, x_n]$ is noetherian.

2.4 Proposition. *If A is noetherian and B is a homomorphic image of A, then B is noetherian. In other words, if A is noetherian, then so is A/I for any ideal I of A.*

2.5 Definition. (a) A ring A is an *(integral) domain* if it has no zero-divisors, i.e., if $xy = 0$ in A then $x = 0$ or $y = 0$.

(b) A is *reduced* if it has no nonzero nilpotent elements, i.e., if $x^n = 0$ for $x \in A$ and $n > 0$, then $x = 0$.

2.6 Definition. The *radical* Rad (I) of an ideal I of A is the ideal of A defined by

$$\text{Rad}(I) = \{x \in A \,|\, x^n \in I \text{ for some } n > 0\} \ .$$

2.7 Definition. A *prime* ideal $\wp$ of A satisfies:

$$\wp \neq A \text{ and if } xy \in \wp, \text{ then } x \in \wp \text{ or } y \in \wp.$$

Remark. $\text{Rad}(\wp) = \wp$ for all prime ideals $\wp$ of A.

2.8 Definition. Let S be a multiplicatively closed subset of A such that $1 \in S$. Then the *ring of fractions* of A with respect to S is

$$S^{-1}A = \left\{ \left[\frac{a}{s}\right] \middle| a \in A \, , \ s \in S \right\}$$

where $\left[\frac{a}{s}\right]$ is an equivalence class of fractions defined as follows:

$$\frac{a_1}{s_1} = \frac{a_2}{s_2} \quad \text{if} \quad (a_1 s_2 - a_2 s_1)t = 0 \text{ for some } t \in S \ .$$

The usual rules of addition and multiplication of fractions make $S^{-1}A$ into a ring.

Let $f \in A$ and $S = \{1, f, f^2, \ldots\}$. Then $S^{-1}A$ is usually written A_f.

Let $\wp$ be a prime ideal. Then $S = A - \wp$ is multiplicatively closed. $S^{-1}A$ is usually written $A_\wp$ and is called "localization of A at $\wp$."

2.9 Definition. An *A-module M* is an abelian group (with the group operation denoted by $+$) on which A acts linearly. That is, if a and b are in A and x, y are in M, then

$$\begin{aligned}
a(x + y) &= ax + ay \\
(a + b)x &= ax + bx \\
(ab)x &= a(bx) \\
1x &= x \ .
\end{aligned}$$

We write ax or xa interchangeably.

Remark. Any ideal I of A is an A-module. In particular, A is an A-module.

Remark. If A is a field, then an A-module is a vector space over A.

2.10 Definition. A *free A-module* M is isomorphic to $\coprod_{j \in J} M_j$, where each M_j is isomorphic to A as an A-module. A *basis* for a free A-module M is a set B of elements of M such that each $u \in M$ can be uniquely written in the form

$$u = \sum_{e \in B} x_e e , \quad x_e \in A ,$$

where all but finitely many $x_e = 0$. If $|B| = n < \infty$, then $M \cong A^n$. If A is noetherian then all bases have the same cardinality, called the *rank* of M.

2.11 Definition. An A-module M is *finitely-generated* if there exist $u_1, \ldots, u_n$ in M such that

$$M = Au_1 + \cdots + Au_n ,$$

i.e., every element u of M can be written (not necessarily uniquely) in the form $u = x_1 u_1 + \cdots + x_n u_n$, $x_i \in A$.

2.12 Proposition. *M can be generated by n elements if and only if M is isomorphic to a quotient of the free A-module A^n.*

Note that an A-module M has one generator (M is called *cyclic*) if and only if it is isomorphic to A/I for some ideal I of A. Although as an A-module A/I is generated by one element (viz., 1), as a *ring* it may require any number of generators.

2.13 Definition. (a) The *annihilator* $\operatorname{Ann} M$ of the A-module M is the ideal of A given by

$$\operatorname{Ann} M = \{x \in A \mid xM = 0\} .$$

(b) An element $x \in A$ is a *non-zero-divisor* on M if whenever $u \in M$ and $xu = 0$, then $u = 0$. In other words, the map $M \xrightarrow{x} M$ given by multiplication by x is injective.

Now let A be a subring of a ring B. We say that B is a *finite A-algebra* if B if finitely-generated as an A-module, i.e., $B = x_1 A + x_2 A + \cdots + x_n A$ for some finite set of elements $x_1, \ldots, x_n$ of B (see Definition 2.11). We also say that B is a *finitely-generated* A-algebra (or of *finite type* over A) if there are finitely many elements $x_1, \ldots, x_n$ of B such that every element of B can be written as a polynomial in $x_1, \ldots, x_n$ with coefficients in A. We then write $B = A[x_1, \ldots, x_n]$. Finally we say that B is *integral* over A if every element of B is the root of a *monic* polynomial whose coefficients belong to A. We then have:

$$\text{finite type} + \text{integral} = \text{finite}.$$

For instance, if B is a finitely-generated algebra over a field $k \subseteq B$ and A is a subalgebra of B, then B is automatically of finite type over A (since B is of finite type over $k \subseteq A$). Thus in this situation, integral=finite.

2.14 Noether Normalization Lemma. *Let k be a field and $A \neq 0$ be a finitely-generated k-algebra. Then there exist elements $y_1, \ldots, y_r \in A$ such that $y_1, \ldots, y_r$ are algebraically independent over k and A is integral over $k[y_1, \ldots, y_r]$.*

A version of this result for graded algebras appears in Chapter 1, Lemma 5.2.

2.15 Definition. Let M_1, M_2, M_3 be A-modules, and

$$f_1 : M_1 \to M_2$$
$$f_2 : M_2 \to M_3$$

be A-module homomorphisms. Then

$$0 \longrightarrow M_1 \xrightarrow{f_1} M_2 \xrightarrow{f_2} M_3 \longrightarrow 0$$

is a *short exact sequence* if f_1 is injective, f_2 is surjective and $\operatorname{im} f_1 = \ker f_2$. (Hence $M_3 \cong M_2/M_1$.)

More generally, define a sequence of A-module homomorphisms

$$\cdots \longrightarrow M_1 \xrightarrow{f_1} M_2 \xrightarrow{f_2} M_3 \xrightarrow{f_3} M_4 \longrightarrow \cdots$$

to be a *(long) exact sequence* if $\operatorname{im} f_i = \ker f_{i+1}$ for all i.

Three useful properties of short exact sequences $0 \to M_1 \to M_2 \to M_3 \to 0$ are the following:

(a) M_2 satisfies ACC if and only if M_1 and M_3 do. Here ACC denotes the *ascending chain condition*: every ascending chain $N_0 \subseteq N_1 \subseteq \cdots$ of A-submodules of an A-module N eventually stabilizes. Note that A itself satisfies ACC (as an A-module) if and only if A is noetherian.

(b) M_2 satisfies DCC if and only if M_1 and M_3 do. Here DCC denotes the *descending chain condition*: every descending chain $N_0 \supseteq N_1 \supseteq \cdots$ of A-submodules of N eventually stabilizes. A module or ring satisfying DCC is called *artinian*.

NOTE: If a ring A (commutative, with identity) satisfies DCC, then it satisfies ACC, but not conversely. However, an A-module may satisfy DCC but not ACC. For instance, the set $M = k[x^{-1}]$ of polynomials in x^{-1} over a field k has an obvious structure as a module over $k[x]$ (set $x^a = 0$ in M for $a > 0$). Then M satisfies DCC but not ACC.

(c) Let M_1, M_2, M_3 be finitely-generated. If $0 \to M_1 \to M_2 \to M_3 \to 0$ is exact, then

$$\text{rank } M_1 - \text{ rank } M_2 + \text{ rank } M_3 = 0 \ ,$$

where rank M is the largest n for which M contains a submodule isomorphic to A^n.

2.16 Definition. Let I be an ideal of the ring A satisfying $\cap I^n = (0)$. We have the quotient maps $p_{n+1} : A/I^{n+1} \longrightarrow A/I^n$, $n \geq 1$. A sequence $(x_1, x_2, \ldots)$ with $x_n \in A/I^n$ is *coherent* if $p_{n+1}x_{n+1} = x_n$ for all $n > 1$. There is an obvious ring structure that can be put on the set of coherent sequences. The resulting ring is the *I-adic completion* of A, denoted $\hat{A}$. We may identify A with the subring of $\hat{A}$ consisting of all eventually constant coherent sequences.

Similarly, if M is an A-module, then we have quotient maps $q_{n+1} : M/I^{n+1}M \longrightarrow M/I^nM$, $n \geq 1$. We define the I-adic completion $\hat{M}$ of M exactly analogously to $\hat{A}$. Then $\hat{M}$ has the natural structure of an $\hat{A}$-module.

For example, if $A = k[x_1, \ldots, x_m]$ and $I = (x_1, \ldots, x_m)$, then $\hat{A} \cong k[[x_1, \ldots, x_m]]$. More generally:

2.17 Proposition. *Let A be noetherian and $I = x_1 A + \cdots + x_m A$. Then*

$$\hat{A} \cong A[[X_1, \ldots, X_m]]/(X_1 - x_1, \ldots, X_m - x_m) \ .$$

2.18 Definition. A *category $\mathcal{C}$* consists of:

(a) A class of *objects $A, B, C, \ldots$*

(b) To each pair of objects A, B of $\mathcal{C}$, a set $\mathcal{C}(A, B)$ of *morphisms* from A to B. If $f \in \mathcal{C}(A, B)$, then write $f : A \to B$.

(c) To each triple, A, B, C of $\mathcal{C}$, a *law of composition*

$$\mathcal{C}(A, B) \times \mathcal{C}(B, C) \to \mathcal{C}(A, C) \ ,$$

where we write $gf = h$ if $(f, g) \mapsto h$, subject to the axioms:

(a_1) The sets $\mathcal{C}(A_1, B_1)$ and $\mathcal{C}(A_2, B_2)$ are disjoint unless $A_1 = A_2$ and $B_1 = B_2$,

(a_2) Given $f : A \to B$, $g : B \to C$, $h : C \to D$, then $h(gf) = (hg)f$.

(a_3) To each object A there is a morphism $1_A : A \to A$ such that for any $f : A \to B$ and $g : C \to A$, we have $f 1_A = f$ and $1_A g = g$.

The main category of concern to us will be $\mathcal{C} = $ **A-Mod**, whose objects are A-modules and whose morphisms are A-module homomorphisms.

2.19 Definition. A (covariant) *functor $F : \mathcal{C} \to \mathcal{D}$* between categories $\mathcal{C}$ and $\mathcal{D}$ is a rule associating with each object X in $\mathcal{C}$ an object FX in $\mathcal{D}$, and with each morphism $f : X \to Y$ in $\mathcal{C}$ a morphism $Ff : FX \to FY$ in $\mathcal{D}$ such that

$$\begin{aligned} F(fg) &= (Ff)(Fg) \\ F1_X &= 1_{FX} \ . \end{aligned}$$

For a *contravariant functor*, we have $Ff : FY \to FX$ and $F(fg) = (Fg)(Ff)$.

2.20 Definition. Let M, N be A-modules. Let C be the free A-module $A^{M \times N}$ (i.e. the free module with one generator for each element of $M \times N$). Let D be the submodule of C generated by all elements of the following types:

$$(x + x', y) - (x, y) - (x', y)$$
$$(x, y + y') - (x, y) - (x, y')$$
$$(ax, y) - a(x, y)$$
$$(x, ay) - a(x, y) .$$

Let $T = C/D$. For each basis element (x, y) of C, let $x \otimes y$ be its image in T. Then T is the *tensor product* of M and N, denoted by $M \otimes N$ or $M \otimes_A N$.

By construction, we have $g : M \times N \to M \otimes N$ such that $g(x, y) := x \otimes y$ is an A-bilinear map. The tensor product satisfies the following universal property: Let P be an A-module. Then if $f : M \otimes N \to P$ is an A-bilinear map, there exists $h : M \times N \to P$, an A-bilinear map, that makes the following diagram commute:

$$M \times N \xrightarrow{\ f\ } P$$

$$g \downarrow \quad \nearrow h$$

$$M \times N$$

If we fix A and M, then the map $F : \mathbf{A\text{-}Mod} \to \mathbf{A\text{-}Mod}$ given by $F(N) = M \otimes_A N$ (or $N \otimes_A M$) is a covariant functor. In other words, given $f : X \to Y$ there is a (canonical) way of defining $M \otimes f : M \otimes X \to M \otimes Y$ satisfying the definition of functor, viz.,

$$(M \otimes f)(u \otimes x) = u \otimes f(x) .$$

2.21 Example. (a) Let $M = N = A = k[x]$, where k is a field (or any ring). Then $M \otimes_k N \cong k[x, y]$, while $M \otimes_A N \cong A$.

(b) For any A-module M and ring A, we have that $M \otimes_A A$ is isomorphic to M via $m \otimes a \mapsto ma$.

(c) If A is an algebra over a field k and K is an extension field of k (so K is a k-vector space), then by "extending the scalars of A to K" we mean forming the K-algebra $A \otimes_k K$.

(d) If S is a multiplicatively closed subset of A containing 1, then we define the *module of fractions* $S^{-1}M$ with respect to S of the A-module M by

$$S^{-1}M = M \otimes_A S^{-1}A .$$

We may regard $S^{-1}M$ either as an A-module or $S^{-1}A$-module.

(e) If I is an ideal of A and $\hat{A}$ the I-adic completion of A, then the I-adic completion $\widehat{M}$ of M is given by

$$\widehat{M} \cong M \otimes_A \hat{A} .$$

2.22 Definition. An A-module P is *projective* if for every *surjective* homomorphism $f : M \to N$ and homomorphism $g : P \to N$, there exists a homomorphism $h : P \to M$ making the following diagram commute:

Equivalently P is projective if every short exact sequence $0 \to L \xrightarrow{\alpha} L' \xrightarrow{\beta} P \longrightarrow 0$ *splits*, i.e., $L' \cong L \oplus P$ such that $\alpha(u) = (u, 0), \beta(u, v) = v$. This turns out to be the same as saying that P is a direct summand of a free module. Hence:

2.23 Proposition. *Free modules are projective.*

2.24 Theorem. *Let $A = k[x_1, \ldots, x_n]$, where k is a field. Then projective A-modules are free.*

NOTE. This theorem, proved independently by Quillen and Suslin, solves a famous problem of Serre. We will only be concerned with *graded-* A-modules (see Chapter 1.2). In this case, the theorem that graded projective modules over $k[x_1, \ldots, x_n]$ are free is much easier to prove and was known from the beginnings of homological algebra.

2.25 Definition. (a) A *chain complex* C over a ring A is a sequence $C = \{C_q, \partial_q\}$ of A-modules C_q and homomorphisms $\partial_q : C_q \to C_{q-1}$ such that $\partial_q \partial_{q+1} = 0$. This is denoted

$$C : \cdots \to C_{q+1} \xrightarrow{\partial_{q+1}} C_q \xrightarrow{\partial_q} C_{q-1} \to \cdots .$$

Since $\partial_q \partial_{q+1} = 0$, we have $\operatorname{im} \partial_{q+1} \subseteq \ker \partial_q$. Define the *q-th homology group* of C by

$$H_q(C) = \ker \partial_q / \operatorname{im} \partial_{q+1} .$$

(b) A *cochain complex* C over A is a sequence $C = \{C_q, \delta_q\}$ of A-modules C_q and homomorphisms $\delta_q : C_q \to C_{q+1}$ such that $\delta_q \delta_{q+1} = 0$. This is denoted

$$C : \cdots \longrightarrow C_{q-1} \xrightarrow{\delta_{q-1}} C_q \xrightarrow{\delta_q} C_{q+1} \longrightarrow \cdots .$$

Define the *q-th cohomology group* of C by

$$H^q(C) = \ker \delta_q / \operatorname{im} \delta_{q-1} .$$

Note that the difference between chain complexes and cochain complexes, and between homology and cohomology, is purely formal. Every cochain complex can be converted to a chain complex by reindexing.

2.26 Definition. A *projective* (respectively, *free*) resolution of an A-module M is an exact sequence

$$\mathcal{P} : \cdots \to P_1 \to P_0 \to M \to 0$$

of projective (respectively, free) A-modules P_i. (It is easily seen that projective resolutions of M always exist.)

2.27 Definition. Let M be an A-module, and let $\mathcal{P}$ be a projective resolution of M as above. If N is another A-module, then we have a chain complex

$$\mathcal{P} \otimes N : \cdots \to P_{n+1} \otimes N \xrightarrow{\partial_{n+1} \otimes 1} P_n \otimes N \xrightarrow{\partial_n \otimes 1} \cdots$$
$$\cdots \to P_0 \otimes N \xrightarrow{\partial_0 \otimes 1} M \otimes N \to 0 .$$

Define

$$\operatorname{Tor}_n^A(M, N) = H_n(\mathcal{P} \otimes N) = \ker(\partial_n \otimes 1) / \operatorname{im}(\partial_{n+1} \otimes 1) .$$

The A-module $\operatorname{Tor}_A^n(M, N)$ does not depend, up to isomorphism, on the choice of projective resolution of M. Moreover, both $\operatorname{Tor}_A^n(M, -)$ and $\operatorname{Tor}_A^n(-, N)$ are covariant functors. Note that $\operatorname{Tor}_0^A(M, N) \cong M \otimes_A N$. A basic property of Tor is the isomorphism $\operatorname{Tor}_n^A(M, N) \cong \operatorname{Tor}_n^A(N, M)$.

2.28 Definition. If M and N are A-modules, then $\mathrm{Hom}_A(M, N)$ denotes the set of all A-module homomorphisms $f : M \to N$. The set $\mathrm{Hom}_A(M, N)$ has the structure of an A-module via

$$(xf)(u) = x(f(u)) ,$$

for $x \in A$, $f \in \mathrm{Hom}_A(M, N)$, $u \in M$. If M and N are free A-modules of ranks m and n, then one can identify in an obvious way (after choosing bases for M and N) elements of $\mathrm{Hom}_A(M, N)$ with $m \times n$ matrices over A. We also set $M^* = \mathrm{Hom}_A(M, A)$. If M is free with basis $u_1, \ldots, u_m$, then M^* is free with dual basis $u_1^*, \ldots, u_m^*$ defined by $u_i^*(u_j) = \delta_{ij}$.

Next we note that $\mathrm{Hom}_A(-, N)$ is a contravariant functor. Namely, if $f : X \to Y$ is a homomorphism of A-modules, then define

$$f^* : \mathrm{Hom}_A(Y, N) \to \mathrm{Hom}_A(X, N)$$

as follows: given $g : Y \to N$ and $u \in X$, let $(f^*g)(u) = gf(u)$.

2.29 Definition. With $\mathcal{P}, M, N$ as in Definition 2.27, we have a cochain complex:

$$\mathrm{Hom}_A(\mathcal{P}, N) : \quad \cdots \leftarrow \mathrm{Hom}(P_{n+1}, N) \xleftarrow{\partial^*_{n+1}} \mathrm{Hom}(P_n, N) \xleftarrow{\partial^*_n}$$

$$\cdots \leftarrow \mathrm{Hom}(P_0, N) \xleftarrow{\partial^*_0} \mathrm{Hom}(M, N) \leftarrow 0 .$$

Define

$$\mathrm{Ext}^n_A(M, N) = H^n(\mathrm{Hom}_A(\mathcal{P}, N)) = \ker \partial^*_{n+1} / \operatorname{im} \partial^*_n .$$

Again, the A-module $\mathrm{Ext}^n_A(M, N)$ does not depend, up to isomorphism, on the choice of $\mathcal{P}$. Moreover, $\mathrm{Ext}^n_A(-, N)$ is a contravariant functor. Note that $\mathrm{Ext}^0_A(M, N) = \mathrm{Hom}_A(M, N)$. (Also, $\mathrm{Ext}^n_A(M, -)$ is a covariant functor, but we will not need this fact.)

2.30 Definition. An A-module I is *injective* if for every homomorphism $f : M \to I$ and *injective* homomorphism $g : M \to N$, there exists a homomorphism $h : N \to I$ making the following diagram commute:

$$
\begin{array}{ccc}
0 \longrightarrow M & \xrightarrow{\ g\ } & N \\
{\scriptstyle f}\downarrow & \nearrow{\scriptstyle h} & \\
I &
\end{array}
$$

In other words, a homomorphism into I can be extended (from M to the larger module N which may be regarded as containing M). Equivalently, I is injective if it is a direct summand of every module which contains it.

2.31 Definition and Theorem. Given an A-module M, there is a unique (up to isomorphism) injective A-module $E_A(M)$ containing M, with the property that every injective A-module containing M also contains $E_A(M)$. $E_A(M)$ is called the *injective hull* or *injective envelope* of M.

2.32 Example. Let $A = k[x_1, \cdots, x_n]$, where k is a field. Regard k as an A-module *via* the isomorphism $k \cong A/(x_1 A + \cdots + x_n A)$. Then $E_A(k) \cong k[x_1^{-1}, \cdots, x_n^{-1}]$. Note that $E_A(k)$ is not finitely-generated.

2.33 Definition. An *injective resolution* of an A-module M is an exact sequence
$$\mathcal{I} : 0 \to M \to I_0 \to I_1 \to \cdots ,$$
where each I_i is an injective A-module. (It is easily seen that injective resolutions of M always exist.)

2.34 Definition. Let $\mathcal{C} = $ **A-Mod**. A functor $F : \mathcal{C} \to \mathcal{C}$ is *additive* if
$$F(M \oplus N) = F(M) \oplus F(N) ,$$
for all objects $M, N \in \mathcal{C}$. A functor F is *left-exact* if for every short exact sequence
$$0 \to M_1 \xrightarrow{f} M_2 \xrightarrow{g} M_3 \to 0 ,$$
the complex
$$0 \to FM_1 \xrightarrow{Ff} FM_2 \xrightarrow{Fg} FM_3$$
is exact. A left-exact functor is additive. An example of a left-exact (covariant) functor is $\mathrm{Hom}_A(N, -)$ for some fixed $N \in \mathcal{C}$.

2.35 Definition. Take an injective resolution of $M \in \mathcal{C} = $ **A-Mod**, say
$$\mathcal{I} : 0 \to M \xrightarrow{\delta_{-1}} I_0 \xrightarrow{\delta_0} I_1 \xrightarrow{\delta_1} \cdots .$$

Let F be a covariant left-exact functor. Applying F to $\mathcal{I}$, we obtain a cochain complex
$$F\mathcal{I} : 0 \to FM \xrightarrow{F\delta_{-1}} FI_0 \xrightarrow{F\delta_0} FI_1 \xrightarrow{F\delta_1} \cdots .$$

The nth *right derived functor* $R^n F$ is defined on objects by
$$R^n F(M) = H^n(F\mathcal{I}) = \ker F\delta_n / \operatorname{im} F\delta_{n-1} .$$

(For $f \in \mathcal{C}(M, N)$ there is a natural definition of $R^n F f$ so that $R^n F$ is indeed a (covariant) functor. We won't define $R^n F f$ here, except to note that if $f : M \to M$ is given by multiplication by $x \in A$, then $R^n F f$ is also given by multiplication by x.) Note that by left-exactness $R^0 F(M) \cong F(M)$.

One could also define the right derived functor of a contravariant functor, and the left derived functors of covariant and contravariant functors (which would include $\text{Tor}_A^n(-, N), \text{Tor}_A^n(M, -)$, and $\text{Ext}_A(-, N)$ as special cases), but we will only need Definition 2.35 here.

2.36 Proposition. *In the setup of the previous definition, let* $0 \to M_1 \xrightarrow{f} M_2 \xrightarrow{g} M_3 \to 0$ *be a short exact sequence in* **A-Mod**. *There is then a long exact sequence*

$$\cdots \longrightarrow R^n F M_1 \xrightarrow{R^n f} R^n F M_2 \xrightarrow{R^n g} R^n F M_3 \xrightarrow{\omega_n} R^{n-1} F M_1 \longrightarrow \cdots$$

$$\cdots \longrightarrow R^1 F M_3 \xrightarrow{\omega_1} R^0 F M_1 \xrightarrow{R^0 f} R^0 F M_2 \xrightarrow{R^0 g} R^0 F M_3 \longrightarrow 0 .$$

(We will not define the homomorphisms ω_n here.)

2.37 Definition. An *augmented chain complex* (over a ring A) is a pair (C, ϵ), where $C = \{C_q, \partial_q\}$ is a chain complex satisfying $C_q = 0$ if $q < 0$, and $\epsilon : C_0 \to A$ is an epimorphism satisfying $\epsilon \partial_1 = 0$. This is denoted

$$(C, \epsilon) : \cdots \xrightarrow{\partial_2} C_1 \xrightarrow{\partial_1} C_0 \xrightarrow{\epsilon} A \longrightarrow 0 .$$

The *reduced homology groups* $\widetilde{H}_q(C)$ (with respect to the augmentation ϵ) are the homology groups of the augmented chain complex (C, ϵ). Because A is a projective (in fact, free) A-module, the epimorphism $\epsilon : C_0 \to A$ splits. It follows that homology and reduced homology are related by

$$H_q(C) \cong \begin{cases} \widetilde{H}_q(C) , & q > 0 \\ \widetilde{H}_0(C) \oplus A , & q = 0 . \end{cases}$$

Note that also $\widetilde{H}_{-1}(C) \cong A$.

Similarly of course we can define augmented cochain complex, reduced cochain complex, and reduced cohomology groups. The only difference is that the monomorphism $\epsilon : A \to C_0$ will not in general split, since A need not be an injective A-module. Of course, if A is a field k, then A is injective and thus $H^0(C) \cong \widetilde{H}^0(C) \oplus k$.

2.38 Definition. Let $C = \{C_q, \partial_q\}$ and $C' = \{C'_q, \partial'_q\}$ be two chain (or cochain) complexes over A. The *tensor product* $C \otimes C' = \{D_n, \epsilon_n\}$ is the chain (or cochain) complex defined by

$$D_n = (C \otimes C')_n = \coprod_{i+j=n} (C_i \otimes C'_j)$$
$$\epsilon_n(C_i \otimes C'_j) = \partial_i C_i \otimes C'_j + (-1)^i C_i \otimes \partial'_j C'_j .$$

The reader should check that $\epsilon_{n-1}\epsilon_n = 0$ (or $\epsilon_n\epsilon_{n-1} = 0$).

For example, in Chapter 1.6 there is considered over a ring R a complex $\otimes_{i=1}^s (0 \to R \to R_{y_i} \to 0)$, where R_{y_i} denotes localization. When $s = 2$ this becomes

$$0 \to R \otimes R \to (R \otimes R_{y_1}) \oplus (R \otimes R_{y_2}) \to R_{y_1} \otimes R_{y_2} \to 0 .$$

Now $R \otimes M \cong R$ and $R_{y_1} \otimes R_{y_2} \cong R_{y_1 y_2}$. Hence we obtain

$$0 \to R \to R_{y_1} \oplus R_{y_2} \to R_{y_1 y_2} \to 0 .$$

2.39. Definition. Let M be an A-module and $x_1, \ldots, x_r \in A$. For $1 \le i \le r$ let Ae_i be a free A-module of rank one with a specified basis e_i. Let $K(x_i)$ denote the chain complex satisfying:

$$K_0(x_i) = A, \ K_1(x_i) = Ae_i$$
$$K_q(x_i) = 0, \text{ if } q \neq 0, 1$$
$$\partial_1(xe_i) = xx_i .$$

This is denoted

$$K(x_i) : 0 \to Ae_i \xrightarrow{x_i} A \to 0 .$$

If M is an A-module, then we have a complex

$$K(x_i, M) = K(x_i) \otimes M : 0 \to Me_i \xrightarrow{x_i} M \to 0 .$$

Define the *Koszul complex* $K(x_1, \ldots, x_r, M)$ of M with respect to $x_1, \ldots, x_r$ by

$$K(x_1, \ldots, x_r, M) = K(x_1, M) \otimes \cdots \otimes K(x_r, M) .$$

If we put $e_{i_1 \cdots i_q} = u_1 \otimes \cdots \otimes u_r$, where $u_i = e_i$ for $i \in \{i_1, \ldots, i_q\}$ and $u_i = 1$ for other i, then $K_q(x_1, \ldots, x_n, M)$ is a free A-module with basis $\{e_{i_1 \cdots i_q} \mid 1 \le i_1 < \cdots < i_q \le r\}$ and thus of rank $\binom{r}{q}$. If $m \in M$, then

$$\partial_q \left(m\, e_{i_1 \cdots i_q} \right) = \sum_{j=1}^{q} (-1)^{j-1} x_{i_j} m\, e_{i_1 \cdots \hat{i}_j \cdots i_q} ,$$

where $\hat{i}_j$ denotes that i_j is missing. This formula is called a *Koszul relation*.

3 Topology

We now give some basic definitions and results from algebraic topology. Any text on algebraic topology should suffice as a reference; for the most part we follow Spanier[145].

An (abstract) *simplicial complex* Δ on a vertex set V is a collection of subsets F of V satisfying:

(a) if $x \in V$ then $\{x\} \in \Delta$,

(b) if $F \in \Delta$ and $G \subset F$, then $G \in \Delta$.

Elements of Δ are called *faces* or *simplices*. If $|F| = q + 1$, then F is a *q-face* or *q-simplex*. We frequently identify the vertex x with the face $\{x\}$.

Suppose V is finite, say $V = \{x_1, \ldots, x_n\}$. Let e_i be the ith unit coordinate vector in $\mathbb{R}^n$. Given a subset $F \subseteq V$, define

$$|F| = \mathrm{cx}\{e_i \mid x_i \in F\}\,,$$

where cx denotes convex hull. Thus if F is an (abstract) q-simplex, then $|F|$ is a geometric q-simplex in $\mathbb{R}^n$. Define the *geometric realization* $|\Delta|$ of the simplicial complex Δ by

$$|\Delta| = \bigcup_{F \in \Delta} |F|\,.$$

Thus $|\Delta|$ inherits from the usual topology on $\mathbb{R}^n$ the structure of a topological space. If X is a topological space homeomorphic to $|\Delta|$, then we (somewhat inaccurately) call Δ a *triangulation* of X. More generally, if $\Gamma = \{\sigma_F : F \in \Delta - \{\emptyset\}\}$ is a set of Euclidean simplices $\sigma_F \subset \mathbb{R}^N$ such that (a) $\dim \sigma_F = |F| - 1$, (b) if $G \subset F$ then σ_G is a face of σ_F, and (c) $\sigma_{F \cap G} = \sigma_F \cap \sigma_G$, then we sometimes call the space $X = \cup \sigma_F$ a geometric realization of Δ.

An *oriented q-simplex* of Δ is a q-simplex F together with an equivalence class of total orderings of F, two orderings being equivalent if they differ by an even permutation of the vertices. Denote by $[v_0, v_1, \ldots, v_q]$ the oriented q-simplex consisting of the q-simplex $F = \{v_0, v_1, \ldots, v_q\}$, together with the equivalence class of orderings containing $v_0 < v_1 < \cdots < v_q$. Fix a ring A (commutative with 1). Let $C_q(\Delta)$ be the free A-module with basis consisting of the oriented q-simplices in Δ, modulo the relations $\sigma_1 + \sigma_2 = 0$ whenever σ_1 and σ_2 are different oriented q-simplices corresponding to the same q-simplex of Δ. Thus $C_q(\Delta) = 0$ for $q < 0$, and for $q \geq 0$ $C_q(\Delta)$ is a free A-module with rank equal to the number of q-simplices of Δ. If Δ is empty, then $C_q(\Delta) = 0$ for all q.

We define homomorphisms $\partial_q : C_q(\Delta) \to C_{q-1}(\Delta)$ for $q \geq 1$ by defining them on the basis elements by

$$\partial_q[v_0, v_1, \ldots, v_q] = \sum_{i=0}^{q}(-1)^i[v_0, v_1, \ldots, \hat{v}_i, \ldots, v_q] \,,$$

where $\hat{v}_i$ denotes that v_i is missing. It is easily verified that ∂_q indeed extends to a homomorphism $C_q(\Delta) \to C_{q-1}(\Delta)$, and that $\partial_q\partial_{q+1} = 0$. The chain complex $C(\Delta) = \{C_q(\Delta), \partial_q\}$ is the *oriented chain complex* of Δ. If $\Delta \neq \emptyset$, then Δ contains $\emptyset$ as a face (of dimension -1). Let $C_{-1}(\Delta)$ be the free A-module with basis $\{\emptyset\}$, and define an augmentation $\epsilon : C_0(\Delta) \to C_{-1}(\Delta) \cong A$ by $\epsilon(x) = \emptyset$ for every vertex $x \in V$. The augmented chain complex $(C(\Delta), \epsilon)$ is the *augmented oriented chain complex* of Δ (over A).

3.1 Definition. The q-*th reduced homology group of* Δ *with coefficients* A, denoted $\widetilde{H}_q(\Delta; A)$, is defined to be the qth homology group of the augmented oriented chain complex of Δ (over A).

Thus the distinction between reduced and ordinary simplicial homology comes about from whether or not we regard the empty set $\emptyset$ as a face.

3.2 Definition. The *reduced Euler characteristic* $\tilde{\chi}(\Delta)$ of Δ is defined by

$$\tilde{\chi}(\Delta) = \sum_{q \geq -1} (-1)^q \operatorname{rank} \widetilde{H}_q(\Delta; A) \,.$$

It is independent of A and is also given by

$$\tilde{\chi}(\Delta) = -1 + f_0 - f_1 + \cdots \,,$$

where f_q is the number of q-simplices in Δ. If $\chi(\Delta)$ is the ordinary Euler characteristic then $\tilde{\chi}(\Delta) = \chi(\Delta) - 1$.

NOTE. If $\Delta \neq \{\emptyset\}$, then $\widetilde{H}_q(\Delta; A) = 0$ for $q < 0$. If $\Delta = \{\emptyset\}$, then

$$\widetilde{H}_q(\{\emptyset\}, A) \cong \begin{cases} A \,, & q = -1 \\ 0 \,, & q \neq -1 \end{cases} \,.$$

In particular, $\tilde{\chi}(\{\emptyset\}) = -1$. If on the other hand $\Delta = \emptyset$, then $\widetilde{H}_q(\Delta; A) = 0$ for all q.

3.3 Proposition. *If* $\Delta \neq \{\emptyset\}$ *or* $\emptyset$, *then* $\widetilde{H}_0(\Delta; A)$ *is a free A-module whose rank is one less than the number of connected components of* Δ.

We now wish to define the homology groups of a space X, rather than a simplicial complex Δ. Let X be a nonempty topological space. Let Δ^q denote the standard q-dimensional ordered geometric simplex $\langle p_0, \ldots, p_q \rangle$ whose vertices p_i are the unit coordinate vectors in $\mathbb{R}^{q+1}$. A *singular q-simplex* in X is a continuous map

$$\sigma : \Delta^q \to X .$$

Let $C_q(X)$ be the free A-module generated by all singular q-simplices. The elements of C_q are formal finite linear combintations $\sum_\sigma c_\sigma \sigma$, where σ is a singular q-simplex and $c_\sigma \in A$. Given a vertex p_i of Δ_q, there is an obvious linear map $e_q^i : \Delta^{q-1} \to \Delta^q$ which sends Δ^{q-1} to the face of Δ^q opposite p_i. The *ith face* of σ, denoted by $\sigma^{(i)}$, is defined to be the singular $(q-1)$-simplex which is the composite

$$\sigma^{(i)} = \sigma \circ e_q^i : \Delta^{q-1} \to \Delta^q \to X .$$

We now define a linear map ($= A$-module homomorphism) $\partial_q : C_q \to C_{q-1}$ by

$$\partial_q(\sigma) = \sum_{i=0}^q (-1)^i \sigma^{(i)} ,$$

where σ is a singular q-simplex. It is easily checked that $\partial_{q-1}\partial_q = 0$, so $C(X) = \{C_q(X), \partial_q\}$ is a chain complex, the *singular chain complex* of X (over A). Define an augmentation $\epsilon : C_0(X) \to A$ by $\epsilon(\sigma) = 1$ for all singular 0-simplices σ. The augmented chain complex $\tilde{C}(X)$ is the *augmented singular chain complex* of X (over A).

3.4 Definition. The *qth reduced singular homology group of X with coefficients A*, denoted $\widetilde{H}_q(X; A)$, is the qth homology group of the augmented singular chain complex of X (over A).

3.5 Definition. The *reduced Euler characteristic* $\tilde{\chi}(X)$ of X is defined by

$$\tilde{\chi}(X) = \sum_{q \geq -1} (-1)^q \operatorname{rank} \widetilde{H}_q(X; A) .$$

It is independent of A.

If Δ is a simplicial complex and Δ_1 and Δ_2 are subcomplexes of Δ, then there is an exact sequence (whose definition we omit)

$$\cdots \to \widetilde{H}_q(\Delta_1 \cap \Delta_2) \to \widetilde{H}_q(\Delta_1) \oplus \widetilde{H}_q(\Delta_2) \to$$
$$\widetilde{H}_q(\Delta_1 \cup \Delta_2) \to \widetilde{H}_{q-1}(\Delta_1 \cap \Delta_2) \to \cdots$$

(with all coefficients A), called the *reduced Mayer-Vietoris sequence of* Δ_1 *and* Δ_2. Similarly, if X is a topological space and X_1, X_2 are "nice" subspaces (e.g., if $X_1 \cup X_2 = (\text{Int}_{X_1 \cup X_2} X_1) \cup (\text{Int}_{X_1 \cup X_2} X_2)$, where $\text{Int}_Y Z$ denotes the relative interior of Z in the space Y), then we have a *reduced Mayer-Vietoris sequence of* X_1 *and* X_2 exactly analogous to that of Δ_1 and Δ_2.

We now come to the relationship between simplicial and singular homology.

3.6 Theorem. *Let Δ be a finite simplicial complex and $X = |\Delta|$. Then there is a (canonical) isomorphism*

$$\widetilde{H}_q(\Delta; A) \cong \widetilde{H}_q(X; A) \, ,$$

for all q.

3.7 Proposition. *Let $\mathbb{S}^{d-1}$ denote a $(d-1)$-dimensional sphere, and let $|\Delta| \approx \mathbb{S}^{d-1}$. Then*

$$\widetilde{H}_q(\Delta; A) \cong \begin{cases} A \, , & q = d - 1 \\ 0 \, , & q \neq d - 1 \end{cases}$$

3.8 Definition. A simplicial complex Δ or topological space X is *acyclic* (over A) if its reduced homology with coefficients A vanishes in all degrees q. (Thus the simplicial complex $\{\emptyset\}$ is not acyclic, since $\widetilde{H}_{-1}(\{\emptyset\}; A) \cong A$.)

3.9 Definition. Let Y be a subspace of X. Then the singular chain module $C_q(Y)$ is a submodule of $C_q(X)$, so we have a quotient complex $C(X, Y) = C(X)/C(Y) = \{C_q(X)/C_q(Y), \bar{\partial}_q\}$. Define the *relative homology* of X modulo Y (with coefficients A) by

$$H_q(X, Y; A) = H_q(C(X, Y)) \, .$$

We next want to define reduced *cohomology* of simplicial complexes and spaces. The simplest way (though not the most geometric) is to dualize the corresponding chain complexes.

3.10 Definition. Let $C'(\Delta) = C(\Delta, \epsilon)$ be the augmented oriented chain complex of the simplicial complex Δ, over the ring A. The *q-th reduced singular cohomology group* of Δ with coefficients A is defined to be

$$\widetilde{H}^q(\Delta; A) = \widetilde{H}^q(\mathrm{Hom}_A(C'(\Delta), A)),$$

where $\mathrm{Hom}_A(C'(\Delta), A)$ is the cochain complex obtained by applying the functor $\mathrm{Hom}_A(-, A)$ to $C'(\Delta)$. Exactly analogously define $\widetilde{H}^q(X; A)$ and $H^q(X, Y; A)$. Sometimes one identifies the free modules $C_q(\Delta)$ and $C^q(\Delta) = \mathrm{Hom}_A(C_q(\Delta), A)$ by identifying the basis of oriented q-chains σ of $C_q(\Delta)$ with its dual basis in $C^q(\Delta)$. Similarly one can identify $C_q(X)$ with $C^q(X)$.

There is a close connection between homology and cohomology of Δ or X arising from the "universal-coefficient theorem for cohomology." We merely mention the (easy) special case that when A is a field k, there are "canonical" isomorphisms

$$\widetilde{H}_q(\Delta; k) \xrightarrow{\cong} \mathrm{Hom}_k(\widetilde{H}^q(\Delta; k), k)$$
$$\widetilde{H}_q(X; k) \xrightarrow{\cong} \mathrm{Hom}_k(\widetilde{H}^q(X; k), k).$$

Thus in particular when $\widetilde{H}_q(\Delta; k)$ is finite-dimensional (e.g., when Δ is finite) we have $\widetilde{H}_q(\Delta; k) \cong \widetilde{H}^q(\Delta; k)$ and similarly for X, but these isomorphisms are not canonical.

3.11 Definition. A topological *n-manifold (without boundary)* is a Hausdorff space in which each point has an open neighborhood homeomorphic to $\mathbb{R}^n$. An *n-manifold with boundary* is a Hausdorff space X in which each point has an open neighborhood which is homeomorphic with $\mathbb{R}^n$ or $\mathbb{R}^n_+ = \{(x_1, \ldots, x_n) \in \mathbb{R}^n \mid x_1 \geq 0\}$. The *boundary* ∂X of X consists of those points with no open neighborhood homeomorphic to $\mathbb{R}^n$. It follows easily that ∂X is either empty or an $(n-1)$-manifold.

Suppose X is a *compact connected n*-manifold with boundary. Then one can show $H_n(X, \partial X; A)$ is either 0 or isomorphic to A.

3.12 Definition. A compact connected *n*-manifold X with boundary is *orientable* (over A) if $H_n(X, \partial X; A) = A$. (The usual definition of orientable is more technical but equivalent to the one given here.)

3.13 Proposition. *Every compact connected n-manifold X with boundary is orientable over a field of characteristic two.*

3.14 Poincaré Duality Theorem. *If a compact connected n-manifold X is orientable over A, then $H_q(X; A) \cong H^{n-q}(X; A)$.*

3.15 Definition. An *n-dimensional pseudomanifold without boundary* (respectively, *with boundary*) is a simplicial complex Δ such that:

(a) Every simplex of Δ is the face of an n-simplex of Δ.

(b) Every $(n-1)$-simplex of Δ is the face of exactly two (respectively, at most two) n-simplices of Δ.

(c) If F and F' are n-simplices of Δ, there is a finite sequence $F = F_1, F_2, \ldots, F_m = F'$ of n-simplices of Δ such that F_i and F_{i+1} have an $(n-1)$-face in common for $1 \le i < m$.

The *boundary* $\partial\Delta$ of a pseudomanifold Δ consists of those faces F contained in some $(n-1)$-simplex of Δ which is the face of exactly one n-simplex of Δ.

3.16 Proposition and Definition. Let Δ be a finite n-dimensional pseudomanifold with boundary. Then either $H_n(\Delta, \partial\Delta; A) \cong A$ or 0. In the former case we say that Δ is *orientable* over A; otherwise *nonorientable*.

3.17 Definition. Let I be the unit interval $[0, 1]$. The *suspension* ΣX of a topological space X is defined to be the quotient space of $X \times I$ in which $X \times 0$ is identified to one point and $X \times 1$ is identified to another point. The *n-fold suspension* $\Sigma^n X$ is defined recursively by $\Sigma^n X = \Sigma(\Sigma^{n-1} X)$.

3.18 Proposition. *For any X and q,*

$$\widetilde{H}_q(X; A) \cong \widetilde{H}_{q+1}(\Sigma X; A) \ .$$

Chapter I

Nonnegative Integral Solutions to Linear Equations

1 Integer stochastic matrices (magic squares)

The first topic will concern the problem of solving linear equations in nonnegative integers. In particular, we will consider the following problem which goes back to MacMahon. Let

$$H_n(r) := \text{ number of } n \times n \text{ N-matrices having line sums } r,$$

where a line is a row or column, and an N-matrix is a matrix whose entries belong to N. Such a matrix is called an *integer stochastic matrix* or *magic square*. Keeping r fixed, one finds that $H_n(0) = 1$, $H_n(1) = n!$, and Anand, Dumir and Gupta [5] showed that

$$\sum_{n \geq 0} \frac{H_n(2)x^n}{(n!)^2} = \frac{e^{x/2}}{\sqrt{1-x}} .$$

See also Stanley [154, Ex. 6.11]. Keeping n fixed, one finds that $H_1(r) = 1$, $H_2(r) = r + 1$, and MacMahon [119, Sect. 407] showed that

$$H_3(r) = \binom{r+4}{4} + \binom{r+3}{4} + \binom{r+2}{4} .$$

Guided by this evidence Anand, Dumir and Gupta [5] formulated the following

1.1 Conjecture. *Fix $n \geq 1$. Then*

(i) $H(r) \in \mathbb{C}[r]$

(ii) $\deg H_n = (n-1)^2$

(iii) $$H_n(-1) = H_n(-2) = \cdots = H_n(-n+1) = 0 \,,$$

$$H_n(-n-r) = (-1)^{n-1} H_n(r) \,.$$

This conjecture can be shown (see [162, Cor. 4.24 and Cor. 4.31]) equivalent to:

$$\sum_{r \geq 0} H_n(r) \lambda^r = \frac{h_0 + h_1 \lambda + \cdots + h_d \lambda^d}{(1-\lambda)^{(n-1)^2+1}} \,, \quad d = n^2 - 3n + 2 \,,$$

$$h_0 + h_1 + \cdots + h_d \neq 0 \,, \quad \text{and} \quad h_i = h_{d-i} \,, \ i = 0, 1, \ldots, d \,.$$

The following additional conjectures can be made:

(iv) $h_i \geq 0$,

(v) $h_0 \leq h_1 \leq \cdots \leq h_{[d/2]}$.

We will verify conjectures (i) to (iv). Conjecture (v) is still open. The solution will appear as a special case of solving linear diophantine equations. This will be done in a ring-theoretic setting, and we will now review the relevant commutative algebra.

2 Graded algebras and modules

Let k be a field, and let R be an $\mathbb{N}^m$-graded connected commutative k-algebra with identity. Thus,

$$R = \coprod_{\alpha \in \mathbb{N}^m} R_\alpha \ (\text{vector space direct sum}) \,, \quad R_\alpha R_\beta \subseteq R_{\alpha+\beta} \,, \quad R_0 = k \,.$$

Elements $x \in R_\alpha$ are said to be *homogeneous* of degree α, denoted $\deg x = \alpha$. Let

$$\mathcal{H}(R) = \bigcup_\alpha R_\alpha \,,$$

and

$$R_+ = \coprod_{\alpha \neq 0} R_\alpha \,.$$

The ideal R_+ (called the *irrelevant ideal*) is maximal; in fact, it is the unique maximal homogeneous ideal.

A $\mathbb{Z}^m$-graded R-module M has a decomposition of the form

$$M = \coprod_{\alpha \in \mathbb{Z}^m} M_\alpha \text{ (vector space direct sum) }, \quad R_\alpha M_\beta \subseteq M_{\alpha+\beta} ,$$

and a map $\phi : M \to N$ between two such modules is *degree-preserving* (or *graded*) if $\phi(M_\alpha) \subseteq N_\alpha$ for all α. As above, write $\mathcal{H}(M) = \cup_\alpha M_\alpha$. A submodule $N \subseteq M$ is *homogeneous* if generated by homogeneous elements, and such a submodule has the structure of a graded R-module by $N_\alpha = N \cap M_\alpha$. In particular, this defines *homogeneous ideal*. If N is a homogeneous submodule of M, then also M/N is a graded R-module:

$$M/N = \coprod_\alpha (M/N)_\alpha , \quad (M/N)_\alpha = M_\alpha/N_\alpha .$$

From now on we assume that R is a finitely-generated k-algebra, i.e., of *finite type* (equivalently, noetherian), and that all graded R-modules are finitely-generated, unless the contrary is explicitly stated. Sometimes it is convenient to be able to consider R-modules as modules over a polynomial ring. Let $y_1, y_2, \ldots, y_s \in \mathcal{H}(R_+)$, $\deg y_i = \delta_i \in \mathbb{N}^m - \{0\}$. Introduce new variables Y_i with $\deg Y_i = \delta_i$, and let $A = k[Y_1, \ldots, Y_s]$. Define a graded A-module structure on R (and hence on any graded R-module) by $Y_i \cdot x = y_i x$ if $x \in R$. R will be a finitely-generated A-module if and only if R is integral over the subring $k[y_1, \ldots, y_s]$. In particular, if $y_1, \ldots, y_s$ generate R then the homomorphism $A \to R$ defined by $Y_i \mapsto y_i$ is surjective, so $R = A/I$ for some homogeneous ideal $I \subseteq A$.

Let M be a finitely-generated $\mathbb{Z}^m$-graded R-module.

2.1 Definition. $H(M, \alpha) = \dim_k M_\alpha < \infty$, $\alpha \in \mathbb{Z}^m$,

$$F(M, \lambda) = \sum_{\alpha \in \mathbb{Z}^m} H(M, \alpha) \lambda^\alpha .$$

$H(M, \alpha)$ is called the *Hilbert function* of M and $F(M, \lambda)$ is called the *Hilbert series* of M. Here $\lambda = (\lambda_1, \lambda_2, \ldots, \lambda_m)$, $\alpha = (\alpha_1, \alpha_2, \ldots, \alpha_m)$, and $\lambda^\alpha = \lambda_1^{\alpha_1} \lambda_2^{\alpha_2} \cdots \lambda_m^{\alpha_m}$. Clearly, $F(M, \lambda) \in \mathbb{Z}[[\lambda_1, \ldots, \lambda_m]] [\lambda_1^{-1}, \ldots, \lambda_m^{-1}]$, since there cannot be arbitrarily large negative exponents due to finite generation.

Given M and $\theta \in R$ let $(0 : \theta) := \{u \in M \mid \theta u = 0\}$. $(0 : \theta)$ is a homogeneous submodule of M. Proofs for the following lemma and theorem can be found in Atiyah–Macdonald [6, Ch. 11] or in Stanley [155, Thm. 3.1].

2.2 Lemma. *Let $\theta \in R_\alpha, \alpha \neq 0$. Then*

$$F(M, \lambda) = \frac{F(M/\theta M, \lambda) - \lambda^\alpha F((0 : \theta), \lambda)}{1 - \lambda^\alpha} .$$

This lemma easily implies the following:

2.3 Theorem. *Suppose R is generated by $y_1, y_2, \ldots, y_s$, $\deg y_i = \delta_i \neq 0$. Then for some $\beta \in \mathbb{Z}^m$ and $P(M, \lambda) \in \mathbb{Z}[\lambda]$,*

$$F(M, \lambda) = \lambda^\beta \frac{P(M, \lambda)}{\prod_{i=1}^s (1 - \lambda^{\delta_i})} .$$

3 Elementary aspects of $\mathbb{N}$-solutions to linear equations

Let us now return to consider $\mathbb{N}$-solutions to linear systems of equations over $\mathbb{Z}$. Many of the details which we omit on this topic may be found in [160]. A more elementary approach appears in [162, Ch. 4.6]. Let Φ be an $r \times n$ $\mathbb{Z}$-matrix, $r \leq n$, and rank $\Phi = r$. Let

$$E_\Phi := \{\beta \in \mathbb{N}^n \,|\, \Phi\beta = \alpha\} ,$$

and for $\alpha \in \mathbb{Z}^r$ let $E_{\Phi,\alpha} := \{\beta \in \mathbb{N}^n \,|\, \Phi\beta = \alpha\}$. E_Φ is clearly a submonoid of $\mathbb{N}^n$, and $E_{\Phi,\alpha}$ is an "E-module," i.e., $E_\Phi + E_{\Phi,\alpha} \subseteq E_{\Phi,\alpha}$.

Let $R_\Phi := kE_\Phi$, the monoid algebra of E_Φ over k. We identify $\beta \in E_\Phi$ with $x^\beta = x_1^{\beta_1} x_2^{\beta_2} \cdots x_n^{\beta_n}$, so that $R_\Phi \subseteq k[x_1, x_2, \ldots, x_n]$ as a subalgebra generated by monomials. Let $M_{\Phi,\alpha} := kE_{\Phi,\alpha}$. Then $M_{\Phi,\alpha}$ is a $\mathbb{Z}^n$-graded R_Φ-module, with grading $\deg x^\beta = \beta$. Clearly,

$$F(R_\Phi, \lambda) = \sum_{\beta \in E_\Phi} \lambda^\beta ,$$

and also

$$F(M_{\Phi,\alpha}, \lambda) = \sum_{\beta \in E_{\Phi,\alpha}} \lambda^\beta .$$

Hence, in this case the ring and module are completely determined by their Hilbert series.

An aside on invariant theory

Let γ_i denote the i-th column of Φ, so $\Phi = [\gamma_1, \gamma_2, \ldots, \gamma_n]$, and let

$$
T = \left\{ \begin{bmatrix} u^{\gamma_1} & & & 0 \\ & u^{\gamma_2} & & \\ & & \ddots & \\ 0 & & & u^{\gamma_n} \end{bmatrix} : u = (u_1, u_2, \ldots, u_r) \in (k^*)^r \right\} .
$$

T is an r-dimensional torus $\subseteq \mathrm{GL}(n, k)$, considered as an algebraic group over k, and T acts on $R = k[x_1, x_2, \ldots, x_n]$ by $\tau x_i = u^{\gamma_i} x_i$, where $\tau = \mathrm{diag}\,(u^{\gamma_1}, u^{\gamma_2}, \ldots, u^{\gamma_n})$. Then

$$
R^T := \{t \in R \mid \tau \cdot t = t, \quad \forall \tau \in T\} = R_\Phi .
$$

Also, for $\alpha \in \mathbb{Z}^r$ and $\tau = \mathrm{diag}\,(u^{\gamma_1}, u^{\gamma_2}, \ldots, u^{\gamma_n}) \in T$, let $\chi_\alpha(\tau) = u^\alpha \in k^*$. χ_α is a one-dimensional character of T and

$$
R^T_{\chi_\alpha} := \{t \in R \mid \tau \cdot t = \chi_\alpha(\tau) \cdot t, \quad \forall \tau \in T\} \cong M_{\Phi,\alpha} .
$$

Thus, some of the developments we present for R_Φ (and $M_{\Phi,\alpha}$) can be seen as special cases of general results about invariant rings of reductive algebraic groups acting on polynomial rings (e.g. Cohen–Macaulayness, due to Hochster and Roberts).

3.1 Theorem. *R_Φ is a finitely-generated k-algebra.*

Proof. Let I be the ideal of $R = k[x_1, x_2, \ldots, x_n]$ generated by $(R_\Phi)_+$. By the Hilbert Basis Theorem I is finitely-generated, i.e., we can find $x^{\delta_1}, x^{\delta_2}, \ldots, x^{\delta_t}$ in $(R_\Phi)_+$ which generate I as an ideal of R.

We want to show that E_Φ is a finitely-generated monoid. *Claim:* $\delta_1, \delta_2, \ldots, \delta_t$ generate E_Φ. Let $\beta \in E_\Phi$. Since $x^\beta \in I$, we get $\beta = \delta_i + \gamma$, $\gamma \in \mathbb{N}^m$. But $\beta, \delta_i \in E_\Phi$ implies $\gamma \in E_\Phi$ (this is the crucial property of this monoid). Having peeled off one generator δ_i, we continue until we get β expressed as a sum of δ_i's. $\qquad\qquad\square$

3.2 Theorem. *$M_{\Phi,\alpha}$ is a finitely-generated R_Φ-module.* The proof is similar.

We now want to find a "smallest" subset $\{\delta_1, \delta_2, \ldots, \delta_t\} \subseteq E_\Phi$ such that R_Φ is a finitely-generated $k[x^{\delta_1}, x^{\delta_2}, \ldots, x^{\delta_t}]$-module.

3.3 Definition. $\beta \in E_\Phi$ is *fundamental* if $\beta = \gamma + \delta$, $\gamma, \delta \in E_\Phi$ implies $\gamma = \beta$ or $\delta = \beta$.

$$\mathrm{FUND}_\Phi := \text{ set of fundamental elements of } E_\Phi .$$

It is clear that FUND_Φ generates E_Φ, and that every set which generates E_Φ contains FUND_Φ. In particular, $|\mathrm{FUND}_\Phi| < \infty$ and

$$R_\Phi = k[x^\delta \,|\, \delta \in \mathrm{FUND}_\Phi] .$$

3.4 Definition. $\beta \in E_\Phi$ is *completely fundamental* if whenever $n > 0$ and $n\beta = \gamma + \delta$ for $\gamma, \delta \in E_\Phi$, then $\gamma = n_1\beta$ for some $0 \le n_1 \le n$.

$$\mathrm{CF}_\Phi := \text{ set of completely fundamental elements of } E_\Phi.$$

3.5 Example. Let $\Phi = [1 \ 1 \ - 2]$, so we are looking for $\mathbb{N}$-solutions to $x + y - 2z = 0$. Then

$$\mathrm{FUND}_\Phi = \{(201), (021), (111)\} , \quad \text{and}$$

$$\mathrm{CF}_\Phi = \{(201), (021)\} , \quad \text{since} \quad 2(111) = (201) + (021) .$$

In the general situation, consider now the set of $\mathbb{R}^+$-solutions β to $\Phi\beta = 0$. It forms a convex polyhedral cone $\mathcal{C}_\Phi$ whose unique vertex is the origin. The integer points nearest 0 on each extreme ray of $\mathcal{C}_\Phi$ form the set CF_Φ. Furthermore, the faces of $\mathcal{C}_\Phi$ (intersections with supporting hyperplanes) are in one-to-one correspondence with $\{\mathrm{supp}\,\beta \,|\, \beta \in E_\Phi\}$, where $\beta = (\beta_1, \beta_2, \ldots, \beta_n) \in \mathbb{N}^n$, $\mathrm{supp}\,\beta = \{i \,|\, \beta_i > 0\}$.

3.6 Example. Consider $x_1 + x_2 - x_3 - x_4 = 0$. The cone of solutions looks like

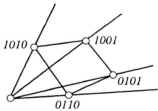

so it has the facial structure of a square. Also, the supports of solutions, ordered by inclusion, yield the face-lattice of a square:

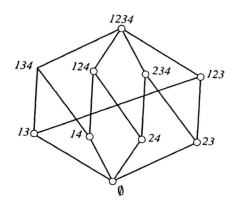

3.7 Theorem. *Let* $\delta_1, \delta_2, \ldots, \delta_t \in E_\Phi$, *and* $S = k[x^{\delta_1}, x^{\delta_2}, \ldots, x^{\delta_t}] \subseteq R_\Phi$. *Then* R_Φ *is a finitely-generated* S-*module (equivalently: integral over* S) *if and only if for every* $\beta \in \mathrm{CF}_\Phi$ *there are* $1 \leq i \leq t$ *and* $j > 0$ *such that* $\delta_i = j\beta$.

Proof. See Proposition 2.2 in Stanley [147]. Visualizing the situation geometrically a proof can be gleaned from the following remarks.

"$\Leftarrow$": If $\delta \in E_\Phi$, then $n\delta = \sum_{\beta \in \mathrm{CF}_\Phi} a_\beta \beta$, which in R_Φ is $(x^\delta)^n = \prod (x^\beta)^{a_\beta}$. Hence R_Φ is integral over $k[x^\beta \mid \beta \in \mathrm{CF}_\Phi]$.

"$\Rightarrow$": If the δ_i's miss some extreme ray $j\beta$, there is no way of reaching any nonzero point on the ray using nonnegative linear combinations. $\square$

3.8 Corollary. *When* $F(R_\Phi, \lambda) = \sum_{\beta \in E_\Phi} \lambda^\beta$ *is written in lowest terms, the denominator is* $\prod_{\beta \in \mathrm{CF}_\Phi} (1 - \lambda^\beta)$.

Proof. Let $\mathrm{CF}_\Phi = \{\beta_1, \beta_2, \ldots, \beta_s\}$, and let $A = k[Y_1, Y_2, \ldots, Y_s]$, deg $Y_i = \beta_i$. R_Φ is an A-module via the multiplication $Y_i \cdot x^\delta = x^{\delta + \beta_i}$. In fact, the previous theorem shows that R_Φ is a finitely-generated A-module. Hence $\prod_{\beta \in \mathrm{CF}_\Phi} (1 - \lambda^\beta)$ is a denominator. To see that it is the *least* denominator, consult [147, Thm. 2.5]. $\square$

In an analogous way it can be shown that either $F(M_{\Phi,\alpha}, \lambda) = 0$ or else has the same denominator as $F(R_\Phi, \lambda)$. The former case occurs, e.g., for $\Phi = [2 \ -2], \alpha = 1$.

4　Integer stochastic matrices again

As an application of the preceding, consider

$$E_\Phi = \left\{(a_{ij})_{i,j=1}^n \,|\, a_{ij} \in \mathbb{N}, \text{ line sums equal}\right\},$$

$$R_\Phi = kE_\Phi = k\left[\prod_{i,j=1}^n x_{ij}^{a_{ij}}\right].$$

The question of which are the completely fundamental elements of E_Φ is in this case answered by the Birkhoff–von Neumann theorem (which says that the extreme points of the convex set of doubly stochastic matrices are the permutation matrices). It follows that every $\alpha \in E_\Phi$ is a sum of permutation matrices, so

$$\text{FUND}_\Phi = CF_\Phi = \{\text{permutation matrices}\}.$$

Thus, $F(R_\Phi, \lambda) = P(\lambda)/\Pi(1 - \lambda^\beta)$, the product taken over all $n \times n$ permutation matrices. Letting $\lambda_{ij} = \begin{cases} x, & i = 1 \\ 0, & i \neq 1 \end{cases}$, so that $\Pi\lambda_{ij}^{a_{ij}} = x^{\text{line sum}}$, we derive

$$\sum_{r \geq 0} H_n(r)x^r = \frac{P_1(x)}{(1 - x)^{n!}},$$

which proves that $H_n(r)$ is a polynomial for large r (cf. the first part of the Anand–Dumir–Gupta conjecture). It will be shown that $H_n(r)$ is a polynomial for *all* r, i.e., that $\deg P_1 < n!$. The substitution $\lambda_{ij} = \begin{cases} x, & i = 1 \\ 0, & i \neq 1 \end{cases}$ is equivalent to "specializing" the $\mathbb{N}^{n^2}$-grading of R_Φ to an $\mathbb{N}$-grading by defining $\deg\left(\Pi\lambda_{ij}^{a_{ij}}\right) = \text{line sum of } (a_{ij}) = \sum_j a_{ij}$.

Now let $E_\Phi := \{n \times n \text{ symmetric } \mathbb{N} \text{ matrices having equal line sums}\}$. In this case FUND_Φ is much harder to describe; fundamental elements with arbitrarily large line-sums can be shown to exist.

4.1 Theorem. *If $\beta \in CF_\Phi$, then the line sum of β is 1 or 2.*

Proof. For $\beta \in CF_\Phi$, by the Birkhoff–von Neumann theorem $\beta = \sum a_i \pi_i$, π_i permutation matrices, $a_i \in \mathbb{N}$. Hence, $2\beta = \beta + \beta^t = \sum a_i(\pi_i + \pi_i^t)$, from which the proof follows. □

Remark. It is possible to characterize the $n \times n$ completely fundamental symmetric matrices, and for $f(n) = |CF_\Phi|$ it can be shown that

$$\sum_{n \geq 0} \frac{f(n)x^n}{n!} = \left(\frac{1 + x}{1 - x}\right)^{1/4} e^{\frac{1}{2}x + \frac{1}{2}x^2}.$$

4.2 Corollary. *Let $S_n(r) := \#\{\beta \in E_\Phi \mid line\ sum\ of\ \beta\ is\ r\}$. Then*

$$\sum_{r \geq 0} S_n(r)x^r = \frac{V(x)}{(1-x)^a(1-x^2)^b}\ , \quad V(x) \in \mathbb{Z}[x]\ , \quad a,b \geq 0\ .$$

Proof. Specialize the grading: $\deg x^\beta$ = line sum of β, and apply Corollary 3.7. □

The form of the generating function reveals that

$$S_n(r) = P_n(r) + (-1)^r Q_n(r)\ , \quad P_n, Q_n \in \mathbb{Q}[r]\ ,$$

for large r, and since $S_n(r) > 0$ for such r, $\deg P_n \geq \deg Q_n$.

5   Dimension, depth, and Cohen–Macaulay modules

Now some more review of commutative algebra. Let R be an $\mathbb{N}^m$-graded k-algebra, and let M be a $\mathbb{Z}^m$-graded R-module. The *Krull dimension* is defined by the following equal numbers:

$\dim R$ = maximum number of (homogeneous) elements of R
       algebraically independent over k
     = length of longest chain of prime ideals of R
     = order to which $\lambda = 1$ is a pole of $F(R, \lambda)$ (if $m = 1$).
$\dim M$ = $\dim(R/\mathrm{Ann}\ M)$
     = order to which $\lambda = 1$ is a pole of $F(M, \lambda)$ (if $m = 1$).

Although the last condition refers to the $m = 1$ case, by specialization of the grading it can apply also in the general case.

5.1 Definition. A *partial h.s.o.p.* (homogeneous system of parameters) for M is a sequence $\theta_1, \theta_2, \ldots, \theta_r \in \mathcal{H}(R_+)$ such that $\dim M/(\theta_1 M + \theta_2 M + \cdots + \theta_r M) = \dim M - r$. An *h.s.o.p.* is a partial h.s.o.p. with $r = \dim M$. Equivalently, $\theta_1, \theta_2, \ldots, \theta_d \in \mathcal{H}(R_+)$ is an h.s.o.p. for $M \Leftrightarrow d = \dim M$ and M is a finitely-generated $k[\theta_1, \theta_2, \ldots, \theta_d]$-module.

Being a partial h.s.o.p. implies being algebraically independent over k, but not conversely. Under what conditions does an h.s.o.p. exist?

5.2 Noether Normalization Lemma. *If $m = 1$ there exists an h.s.o.p. for M. If $|k| = \infty$ and R is generated by R_1 (or even that R is integral over the subalgebra $k[R_1]$ generated by R_1), then we can choose an h.s.o.p. from R_1.*

The theorem fails for $m > 1$, e.g., $R = k[x, y]/(xy)$ with $\deg x = (1, 0)$, $\deg y = (0, 1)$ lacks an h.s.o.p..

Let us return momentarily to the ring R_Φ. Using that monomials $x^{\beta_1}, x^{\beta_2}, \ldots, x^{\beta_t}$, for $\beta_i \in E_\Phi$, are algebraically independent over k if and only if the vectors $\beta_1, \beta_2, \ldots, \beta_t$ are linearly independent over $\mathbb{Q}$, we find:

$$\begin{aligned}
\dim R_\Phi \quad &= \text{maximum number of linearly independent elements in } E_\Phi \\
&= \text{dimension of vector space spanned by } E_\Phi \text{ over } \mathbb{Q} \\
&= n - \text{rank}(\Phi) \text{ (assuming that } \exists \beta > 0, \, \beta \in E_\Phi).
\end{aligned}$$

Now recall the polynomials (for r large) $H_n(r)$ and $P_n(r)$ related to the enumeration of $n \times n$ (symmetric) $\mathbb{N}$-matrices with constant line-sum r.

5.3 Corollary.

(i) $\deg H_n(r) = (n - 1)^2$,

(ii) $\deg P_n(r) = \binom{n}{2}$.

Proof. The degree of these polynomials is one less than the order to which $\lambda = 1$ is a pole of $F(R_\Phi, \lambda)$. In the first case Φ is an $(2n - 2) \times n^2$-matrix and in the second case an $(n - 1) \times \binom{n+1}{2}$-matrix, so the proof follows from the above description of $\dim R_\Phi$. $\qquad\square$

Recall that we had $S_n(r) = P_n(r) + (-1)^r Q_n(r)$, with $\deg Q_n \leq \deg P_n = \binom{n}{2}$. Concerning the problem of finding $\deg Q_n$ the following can be proved.

5.4 Theorem. *Let $E_\Phi := \{n \times n$ symmetric $\mathbb{N}$-matrices with equal line sums$\}$, $\deg x^\beta := $ line sum of β. Let*

$$\begin{aligned}
f_n &:= \min\{j \mid \theta_1, \theta_2, \ldots, \theta_d \text{ is an h.s.o.p. for } R_\theta , \\
&\deg \theta_1 = \cdots = \deg \theta_j = 2 , \ \deg \theta_{j+1} = \cdots = \deg \theta_d = 1\} .
\end{aligned}$$

Then

$$f_n = \begin{cases} \binom{n-1}{2} , & n \text{ odd} \\ \binom{n-2}{2} , & n \text{ even} . \end{cases}$$

5.5 Theorem. $\deg Q_n = f_n - 1$.

Proof. Stanley [151, Prop. 5.4] showed that $\deg Q_n \leq f_n - 1$ and conjectured that equality holds. This conjecture was proved by Jia [103], [104].
$\square$

5.6 Definition. $\theta_1, \theta_2, \ldots, \theta_r \in \mathcal{H}(R_+)$ is a *homogeneous M-sequence (regular sequence)* if θ_{i+1} is a non-zero-divisor on $M/(\theta_1 M + \cdots + \theta_i M)$, $0 \leq i < r$. Equivalently, $\theta_1, \theta_2, \ldots, \theta_r$ are algebraically independent over k and M is a *free* $k[\theta_1, \theta_2, \ldots, \theta_r]$-module.

An M-sequence is a partial h.s.o.p., and if $m = 1$ any two maximal M-sequences have the same length. The latter is not true for $m > 1$, e.g. letting $R = k[x, y, z]/(xy - z^2)$, $\deg x = (2, 0)$, $\deg y = (0, 2)$, and $\deg z = (1, 1)$, then $\{x, y\}$ and $\{z\}$ are maximal homogeneous R-sequences. In terms of the Hilbert series we get the following characterization: $\theta_1, \theta_2, \ldots, \theta_r \in \mathcal{H}(R_+)$ is an M-sequence if and only if

$$F(M, \lambda) = \frac{F(M/(\theta_1 M + \cdots + \theta_r M), \lambda)}{\prod_{i=1}^{r}(1 - \lambda^{\deg \theta_i})}.$$

5.7 Definition. (i) If $m = 1$, let *depth* $M :=$ length of longest homogeneous M-sequence.

(ii) If $m > 1$, specialize the grading to a $\mathbb{Z}$-grading in any way and define *depth* M as in (i). (It can be shown that this definition is independent of the specialization.)

It is clear that depth $M \leq \dim M$. The case of equality, i.e., when some h.s.o.p. is regular, is of particular importance.

5.8 Definition. M is *Cohen–Macaulay* if depth $M = \dim M$.

5.9 Theorem. *Let M have an h.s.o.p. Then M is Cohen–Macaulay*

$\Leftrightarrow$ *every h.s.o.p. is regular*
$\Leftrightarrow$ *M is a finitely-generated and free $k[\theta]$-module for some (equivalently, every) h.s.o.p. $\theta = (\theta_1, \theta_2, \ldots, \theta_d)$.*

5.10 Theorem. *Let M be Cohen–Macaulay, with an h.s.o.p. $\theta = (\theta_1, \theta_2, \ldots, \theta_d)$. Let $\eta_1, \eta_2, \ldots, \eta_t \in \mathcal{H}(M)$. Then $M = \coprod_{i=1}^{t} \eta_i k[\theta]$ if and only if $\eta_1, \eta_2, \ldots, \eta_t$ is a k-basis for $M/\theta M$. For such a choice of θ's and η's it follows that*

$$F(M, \lambda) = \frac{\sum_{i=1}^{t} \lambda^{\deg \eta_i}}{\prod_{j=1}^{d}(1 - \lambda^{\deg \theta_j})}.$$

Returning to our ring R_Φ once more, we can now state the following theorem, which will be proved later.

5.11 Theorem(Hochster [94]). *R_Φ is Cohen–Macaulay.*

5.12 Corollary.

$$\sum_{r \geq 0} H_n(r)\lambda^r = \frac{P(\lambda)}{(1-\lambda)^{(n-1)^2+1}} , \quad P(\lambda) \in \mathbb{N}[\lambda] .$$

The point of Corollary 5.12 is that the polynomial $P(\lambda)$ has *nonneg-ative* coefficients. The corollary follows since permutation matrices have degree one. It is an open problem to compute $P(\lambda)$ or even $P(1)$ in a simple way. In particular, can $P(1)$ be computed more quickly than $P(\lambda)$? For some work related to the problem of computing $P(\lambda)$, see [63], [64].

5.13 Theorem. *Let* $\dim R_\Phi = d$. *There exist free commutative monoids* $G_1, G_2, \ldots, G_t \subseteq E_\Phi$, *all of rank d, and also* $\eta_1, \eta_2, \ldots, \eta_t \in E_\Phi$, *such that* $E_\Phi = \biguplus_{i=1}^t (\eta_i + G_i)$, *where* $\biguplus$ *denotes disjoint union.*

In terms of the ring this theorem says that $R_\Phi = \coprod_{i=1}^t x^{\eta_i} k[G_i]$. This is analogous to the Cohen–Macaulay property, but differs in that the G_i's change. The proof is combinatorial, and uses the shellability of convex polytopes (due to Bruggesser and Mani). The proof is sketched in [160, §5]. In [160, Conj. 5.1] there is a conjectured generalization for any Cohen-Macaulay $\mathbb{N}^m$-graded algebra.

5.14 Example. For the equation $x_1 + x_2 - x_3 - x_4 = 0$ we get

$$R_\Phi = k[x_1 x_3, x_1 x_4, x_2 x_3] \oplus x_2 x_4 k[x_1 x_4, x_2 x_3, x_2 x_4] .$$

Here $x_1 x_3$ corresponds to the solution $(1, 0, 1, 0)$ as usual, and the geometry of the cone of solutions after triangulation is

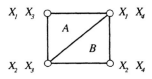

Geometrically, we have taken all integer points in cone A, and "pushed off" cone B from its intersection with A by translation by $(0, 1, 0, 1)$.

6   Local cohomology

We now turn to the proof that R_Φ is Cohen–Macaulay, and more generally to the question of deciding depth $M_{\Phi,\alpha}$. For this we shall use the tool of local cohomology, which will first be reviewed. As always, all rings R and modules M are graded. Local cohomology will be defined with respect to the irrelevant ideal R_+.

Let $L_{R_+}(M) = L(M) = \{u \in M \mid R_+^n u = 0 \text{ for some } n > 0\}$. If $f : M \to N$, then $L(f) : L(M) \to L(N)$ by restriction. It is easy to check that L is a left-exact additive functor, so we can take the right-derived functors $R^i L$.

6.1 Definition. $H^i(M) = H^i_{R_+}(M) = R^i L_{R_+}(M)$.

Some of the fundamental properties of local cohomology $H^i(M)$ will now be stated. (A good general reference is [48].) In particular, the reader unfamiliar or unenamored with homological algebra can adopt the following theorem as the definition of $H^i(M)$. In the following R_{y_i} denotes R localized at y_i (i.e., with respect to the multiplicative set generated by y_i), and $M_{y_i} = M \otimes_R R_{y_i}$.

6.2 Theorem.

$$H^i(M) = H^i \left[\bigotimes_{i=1}^{s} (0 \to R \to R_{y_i} \to 0) \otimes M \right]$$

$$= H^i \left[0 \xrightarrow{\delta_0} M \xrightarrow{\delta_1} \coprod_i M_{y_i} \xrightarrow{\delta_2} \coprod_{i<j} M_{y_i y_j} \xrightarrow{\delta_3} \cdots \xrightarrow{\delta_s} M_{y_1 \cdots y_s} \to 0 \right]$$

$$= \ker \delta_{i+1} / \operatorname{im} \delta_i \,,$$

where $y_1, y_2, \ldots, y_s \in \mathcal{H}(R_+)$ *and* $\operatorname{Rad}(y_1, \ldots, y_s) = R_+$.

The complex $\otimes_{i=1}^{s}(0 \to R \to R_{y_i} \to 0) \otimes M$ is denoted $\mathcal{K}(\mathbf{y}^\infty, M)$. Local cohomology is depth sensitive in the following sense.

6.3 Theorem. $H^i(M) = 0$ *unless* $e = \operatorname{depth} M \le i \le \dim M = d$; *and* $H^e(M) \ne 0$, $H^d(M) \ne 0$.

Theorem 6.2 imposes on $H^i(M)$ in a natural way the structure of a $\mathbb{Z}^m$-graded module: $H^i(M) = \coprod_{\alpha \in \mathbb{Z}^m} H^i(M)_\alpha$. Since $H^i(M)$ is known to be artinian, it follows that $H^i(M)_\alpha = 0$ for $\alpha \gg 0$. However, $H^i(M)$ is usually not finitely-generated. Define

$$F\left(H^i(M), \lambda\right) = \sum_{\alpha \in \mathbb{Z}^m} \left(\dim_k H^i(M)_\alpha\right) \lambda^\alpha \,.$$

6.4 Theorem. $F(M, \lambda)_\infty = \sum_{i=e}^{d} (-1)^i F(H^i(M), \lambda)$.

In this formula $F(M, \lambda)_\infty$ signifies that $F(M, \lambda)$ is to be expanded as a Laurent series around ∞. For instance, for $R = k[x]$, $H^1(R) = x^{-1}k[x^{-1}]$ and $F(R, \lambda) = \frac{1}{1-\lambda} = -\frac{\lambda^{-1}}{1-\lambda^{-1}} = -\sum_{n<0} \lambda^n = -F(H^1(R), \lambda)$. Let $\deg F(R, \lambda)$ denote the degree as a rational function, i.e., degree of numerator minus degree of denominator.

6.5 Corollary. *If $m = 1$ and $H^i(R)_\alpha = 0$ for all $\alpha > 0$ and all i, then $\deg F(R, \lambda) < 0$.*

Proof. Write (uniquely) $F(R, \lambda) = G(R, \lambda) + L(R, \lambda)$ where $\deg G(R, \lambda) < 0$ and $L(R, \lambda) \in \mathbb{Z}[\lambda]$. The expansion $G(R, \lambda)_\infty$ has only negative exponents, while $L(R, \lambda)_\infty = L(R, \lambda)$. Hence if $H^i(R)_\alpha = 0$ for $\alpha > 0$ and all i, then by the previous theorem $F(R, \lambda) = G(R, \lambda)$.   □

The condition $\deg(R, \lambda) < 0$ is equivalent to saying that the Hilbert function $H(R, n)$ has no "exceptional" values. E.g., if R is generated by R_1 then $H(R, n)$ is a polynomial for *all* $n \geq 0$.

7 Local cohomology of the modules $M_{\Phi, \alpha}$

Let Φ be an $r \times n$ $\mathbb{Z}$-matrix and $\alpha \in \mathbb{Z}^n$ as before, and recall the definitions of $E_\Phi, E_{\Phi,\alpha}, R_\Phi$ and $M_{\Phi,\alpha}$. Let $\overline{E}_\Phi$ denote the group generated by E_Φ in $\mathbb{Z}^n$, and let $\overline{E}_{\Phi,\alpha} := \overline{E}_\Phi + E_{\Phi,\alpha}$, the coset of $\overline{E}_\Phi$ in $\mathbb{Z}^n$ containing $E_{\Phi,\alpha}$. The nonnegative real solutions to $\Phi\beta = 0$ form a convex polyhedral cone $\mathcal{C}_\Phi = \{\beta \in (\mathbb{R}^+)^n \mid \Phi\beta = 0\}$. Any cross-section of $\mathcal{C}_\Phi$, i.e., bounded intersection with a hyperplane meeting the relative interior of $\mathcal{C}_\Phi$, is a convex polytope, and different cross-sections are combinatorially equivalent. If F is a face of the cross-section polytope $\mathcal{P}_\Phi$, define $\operatorname{supp} F := \operatorname{supp} \beta$ for any $\beta \in F - \partial F$, where as before $\operatorname{supp}(\beta_1, \ldots, \beta_n) = \{i \mid \beta_i > 0\}$.

We now try to compute the local cohomology of R_Φ by considering the complex $\mathcal{K}(\mathbf{y}^\infty, M)$, $M = M_{\Phi,\alpha}$. Recall the set of completely fundamental solutions $\mathrm{CF}_\Phi = \{\delta_1, \delta_2, \ldots, \delta_s\}$, consisting of the integer points nearest the origin on the extreme rays of $\mathcal{C}_\Phi$. Thus, CF_Φ is in one-to-one correspondence with the vertices of the polytope $\mathcal{P}_\Phi$. Set $y_i = x^{\delta_i} \in R_\Phi$. We know by Theorem 3.7 that $\mathrm{Rad}(y_1, \ldots, y_s) = R_+$. For $\beta \in \overline{E}_{\Phi,\alpha}$ consider the part of $\mathcal{K}(\mathbf{y}^\infty, M)$ of degree β :

$$0 \to M_\beta \to \coprod_i (M_{y_i})_\beta \to \coprod_{i<j} \left(M_{y_i y_j}\right)_\beta \to \cdots . \tag{1}$$

In, say, $M_{y_i y_j}$ we have inverted everything on the face spanned by y_i and y_j. So, to get something of degree β in $M_{y_i y_j}$ take anything in $E_{\Phi,\alpha}$

and subtract δ_i and δ_j any number of times (i.e., multiply by $x^{-\delta_i}$ and $x^{-\delta_j}$). In general, there is something in $M_{y_{i_1}\cdots y_{i_r}}$ of degree β if and only if $N(\delta_{i_1} + \cdots + \delta_{i_r}) + \beta \geq 0$ for $N \gg 0$. This condition is equivalent to $\mathrm{supp}_- \beta \subseteq \mathrm{supp}\, F$, where F is the face spanned by $\delta_{i_1}, \ldots, \delta_{i_r}$ and $\mathrm{supp}_-(\beta_1, \ldots, \beta_n) = \{ i \mid \beta_i < 0 \}$, the *negative support* of β. Thus

$$\dim_k \left(M_{y_{i_1}} \cdots y_{i_r} \right)_\beta = \begin{cases} 1, & \text{if } \mathrm{supp}_- \beta \subseteq \mathrm{supp}\, F \\ 0, & \text{otherwise .} \end{cases}$$

So all the pieces of the β-part (1) of $K(\mathbf{y}^\infty, M)$ are 0- or 1-dimensional vector spaces. Now, the key fact is that we can identify the complex (1) with the augmented chain complex of the simplicial complex Δ_β whose faces are the sets $S \subseteq \mathrm{CF}_\Phi$ such that

$$\begin{aligned} \mathrm{supp}_- \beta \; &\subseteq \; \bigcup_{\delta \in \mathrm{CF}_\Phi - S} (\mathrm{supp}\, \delta) \\ &= \; \mathrm{supp}\, (\text{face } F \text{ of } \mathcal{P}_\Phi \text{ spanned by} \\ &\qquad \text{all } \delta \in \mathrm{CF}_\Phi - S). \end{aligned}$$

We may therefore compute local cohomology by computing reduced simplicial homology. For further details, see [160].

7.1 Example. Consider the equation $x_1 + x_2 - x_3 - x_4 = 2$, i.e., let $\Phi = [1\ 1\ -1\ -1]$ and $\alpha = 2$. Here $\mathrm{FUND}_\Phi = \mathrm{CF}_\Phi$ consists of the four elements $(1,0,1,0), (1,0,0,1), (0,1,0,1), (0,1,1,0)$, so $y_1 = x_1 x_3$, $y_2 = x_1 x_4$, $y_3 = x_2 x_4$ and $y_4 = x_2 x_3$. Geometrically, $\mathcal{P}_\Phi$ is a square. Let $\beta = (0,0,-1,-1)$ and $M = M_{\Phi,\alpha}$, and write $M_i = (M_{y_i})_\beta$, $M_{ij} = \left(M_{y_i y_j} \right)_\beta$, and so on. Then the complex (1) takes the following form, where the non-zero pieces are underlined:

$$\begin{aligned} 0 \; &\rightarrow \; M \rightarrow M_1 \oplus M_2 \oplus M_3 \oplus M_4 \rightarrow \\ &\rightarrow \; \underline{M_{12}} \oplus \underline{M_{13}} \oplus M_{14} \oplus M_{23} \oplus \underline{M_{24}} \oplus \underline{M_{34}} \rightarrow \qquad (2) \\ &\rightarrow \; \underline{M_{123}} \oplus \underline{M_{124}} \oplus \underline{M_{134}} \oplus M_{234} \rightarrow M_{1234} \rightarrow 0 \,. \end{aligned}$$

Taking complements of the underlined elements we find that

$$\Delta_\beta = \{ 34, 24, 13, 12, 4, 3, 2, 1, \emptyset \} \,,$$

and the homology of Δ_β (a circle) is non-zero only in dimension one, corresponding to the $M_{12} \oplus \cdots$ part of (2). In terms of local cohomology this means that $\dim_k (H^2(M_{\Phi,\alpha})_\beta) = 1$, so $M_{\Phi,\alpha}$ is not Cohen–Macaulay, since $\dim M_{\Phi,\alpha} = 3$.

Let $\mathcal{P}^*$ be the dual polytope of $\mathcal{P} = \mathcal{P}_\Phi$, and define $\Gamma_\beta = \cup F^*$, the union over all faces F of $\mathcal{P}$ such that $\mathrm{supp}_-\,\beta \subseteq \mathrm{supp}\,F$, where F^* in the face of $\mathcal{P}^*$ corresponding to F under the order-reversing bijection between the face lattices of $\mathcal{P}$ and $\mathcal{P}^*$. Thus, Γ_β is a polyhedral subcomplex of $\mathcal{P}^*$. Let $d = \dim R_\Phi$, so $\dim \mathcal{P} = d - 1$, and let $s = |\mathrm{CF}_\Phi|$. Our next result show that we can replace the simplicial complex Δ_β by the more tractable Γ_β.

7.2 Theorem. $\widetilde{H}_i(\Gamma_\beta; k) \cong \widetilde{H}_{s-d+i}(\Delta_\beta; k)$.

The proof uses Alexander duality twice and also a theorem on lattice homology of J. Folkman [76]. See [160, Lemma 2.8]. In view of this result we conjecture that Δ_β has the homotopy type of the $(s - d)$-fold suspension of Γ_β.

7.3 Theorem. *Let* $M = M_{\Phi,\alpha}$, $d = \dim M$. *Then as vector spaces over* k,

$$H^i(M)_\beta \cong \begin{cases} \widetilde{H}_{d-1-i}(\Gamma_\beta; k)\,, & \text{if } \beta \in \overline{E}_{\Phi,\alpha} \\ 0\,, & \text{otherwise}\,. \end{cases}$$

Keeping in mind that $\widetilde{H}_i(\emptyset; k) = k$ if $i = -1$ and $= 0$ otherwise, we deduce the following.

7.4 Corollary. *M is Cohen–Macaulay if and only if for all $\beta \in \overline{E}_{\Phi,\alpha}$ either $\Gamma_\beta = \emptyset$ or Γ_β is acyclic.*

7.5 Corollary. $H^d(M)_\beta \cong \begin{cases} k\,, & \text{if } \beta \in \overline{E}_{\Phi,\alpha} \text{ and } \Gamma_\beta = \emptyset \\ 0\,, & \text{otherwise}\,. \end{cases}$

For the Cohen–Macaulay condition above note that $\Gamma_\beta = \emptyset$ if and only if $\mathrm{supp}_-\,\beta \subseteq \mathrm{supp}\,F$ implies $F = \mathcal{P}$. Also, if $\alpha = 0$ then $\Gamma_\beta = \emptyset \Leftrightarrow \mathrm{supp}_-\,\beta = \mathrm{supp}\,\mathcal{P}$ (there is a simple geometric proof for this). In particular, if there exists $\gamma \in E_\Phi$ such that $\gamma > 0$, then for $\beta \in \overline{E}_\Phi : \Gamma_\beta = \emptyset \Leftrightarrow \beta < 0$ (strict inequality in all coordinates). Hence, for $\alpha = 0$ there is the following "reciprocity" result (when $E_\Phi \cap \mathbb{P}^n \neq \emptyset$):

R_Φ is spanned by *nonnegative* solutions to $\Phi\beta = 0$, and

$H^d(R_\Phi)$ is spanned by *strictly negative* solutions to $\Phi\beta = 0$.

In general, $H^d(M_{\Phi,\alpha}) \cong k\{x^\beta \mid \beta \in \overline{E}_{\Phi,\alpha}, \Gamma_\beta = \emptyset\}$, with the R_Φ-module structure given by

$$x^\gamma \cdot x^\beta = \begin{cases} x^{\gamma+\beta}\,, & \text{if } \Gamma_{\gamma+\beta} = \emptyset \\ 0\,, & \text{otherwise}\,, \end{cases}$$

for $\gamma \in E_\Phi, x^\beta \in H^d(M_{\Phi,\alpha})$.

7.6 Corollary. *(a) (Hochster [94]) R_Φ is Cohen–Macaulay, and (b) for any specialization of R_Φ to an $\mathbb{N}$-grading, $\deg F(R_\Phi, \gamma) < 0$.*

Proof. (a) We must show that if $\beta \in \overline{E}_\Phi$ and $\Gamma_\beta \neq \emptyset$ then Γ_β is acyclic. Assume that β satisfies these hypotheses. We will sketch an argument that Γ_β is in fact a ball. We may assume $\beta \notin E_\Phi$, since if $\beta \in E_\Phi$ then $\Gamma_\beta = \mathcal{P}^*$. Stand at point β in the d-dimensional vector space spanned by the cone C_Φ, and look at $\mathcal{P}$. It is not hard to show, since $\Gamma_\beta \neq \emptyset$, that the faces of $\mathcal{P}$ visible from β (regarding C_Φ as opaque) form a $(d-2)$-ball. The faces F you don't see are those such that $\mathrm{supp}_- \beta \subseteq \mathrm{supp}\, F$. Passing to the dual polytope $\mathcal{P}^*$, the non-visible part of $\mathcal{P}$ goes to Γ_β. By polytopal duality Γ_β is a ball (and hence acyclic).

(b) Part (a) gives that $F(R_\Phi, \lambda)_\infty = (-1)^d F(H^d(R_\Phi), \lambda)$. Thus, $F(R_\Phi, \lambda)_\infty$ has no exponents ≥ 0, so as before we conclude that the Hilbert function of R_Φ has no exceptional values. $\square$

Thus we have also proved Corollary 5.12. An analogue of Corollary 5.12 for symmetric magic squares follows in the same way.

While the modules $M_{\Phi,\alpha}$ may fail to be Cohen–Macaulay in general, some of them which are "close enough" to R_Φ are in fact Cohen–Macaulay. Let us mention some results in this direction.

7.7 Corollary. *Suppose there exists $\gamma = (\gamma_1, \ldots, \gamma_n) \in \mathbb{Q}^n$, $-1 < \gamma_i \leq 0$, such that $\Phi\gamma = \alpha$. Then $M_{\Phi,\alpha}$ is Cohen–Macaulay.*

Proof. Let $\Phi\beta = \alpha$, $\beta \in \overline{E}_{\Phi,\alpha}$. Then $\Phi(\beta - \gamma) = 0$ and $\mathrm{supp}_- \beta = \mathrm{supp}_-(\beta - \gamma)$. Γ_β depends only on $\mathrm{supp}_- \beta$, so $\Gamma_\beta = \Gamma_{m(\beta-\gamma)}$, where m is chosen so that $m(\beta - \gamma)$ is integral. Since R_Φ is Cohen–Macaulay the complex $\Gamma_{m(\beta-\gamma)}$ is empty or acyclic. Hence, so is Γ_β and we are done. $\square$

As an aside we mention that given $\Phi = [\rho_1 \rho_2 \cdots \rho_n]$ the number of α's satisfying the hypotheses of the previous corollary equals

$$\sum_{\substack{\rho_{i_1} \cdots \rho_{i_j} \\ \text{lin. indep}}} (-1)^{r-j} \ \mathrm{g.c.d.} \left(j \times j\text{-minors of } \left[\rho_{i_1} \cdots \rho_{i_j} \right] \right) .$$

7.8 Corollary. *Let $\Phi = [a_1, \ldots, a_s, -b_1, \ldots, -b_t]$, $a_i, b_j > 0$, $s, t > 0$. Thus, $d = \dim R_\Phi = s+t-1$. Let $\Phi \begin{bmatrix} x \\ y \end{bmatrix} = \alpha$ stand for $\sum a_i x_i - \sum b_j y_j = \alpha$. Then for $0 \leq i < d$:*

$$H^i(M_{\Phi,\alpha}) \cong \begin{cases} k, & \text{if } i = s, \text{ and } \Phi \begin{bmatrix} x \\ y \end{bmatrix} = \alpha \text{ for some } x < 0 \text{ and } y \geq 0 \\ k, & \text{if } i = t, \text{ and } \Phi \begin{bmatrix} x \\ y \end{bmatrix} = \alpha \text{ for some } x \geq 0 \text{ and } y < 0 \\ 0, & \text{otherwise} . \end{cases}$$

Consequently,

$$\text{depth } M_{\Phi,\alpha} = \begin{cases} 0 \,, & \text{if } M_{\Phi,\alpha} = 0 \\ s \,, & \text{if } \exists \text{ solution } x < 0,\ y \geq 0 \\ t \,, & \text{if } \exists \text{ solution } x \geq 0,\ y < 0 \\ s + t - 1 \,, & \text{otherwise} \,. \end{cases}$$

The proof is based on topology — it is possible to compute the complexes Γ_β explicitly. Note that the middle two conditions are mutually exclusive; if there exists a solution $x < 0, y \geq 0$ then $\alpha < 0$ and in the other case $\alpha > 0$.

In conclusion we mention the following result, though no applications of it have yet been found. It was discovered independently by M. Hochster, the author, and perhaps others.

7.9 Theorem. *Let $E \subseteq \mathbb{N}^m$ be a finitely-generated monoid. Let $G \subseteq \mathbb{Z}^m$ be a finitely-generated "E-module," i.e., $E + G \subseteq G$. Let $R = kE$ and $M = kG$. Let $V = \{\beta_1, \beta_2, \ldots, \beta_t\} \subseteq E$ such that R is integral over $k\left[x^{\beta_1}, x^{\beta_2}, \ldots, x^{\beta_t}\right]$ (the subalgebra generated by the x^{β_i}'s). Let $A = k[y_1, y_2, \ldots, y_t] \to k\left[x^{\beta_1}, x^{\beta_2}, \ldots, x^{\beta_t}\right]$ be the surjection given by $y_i \mapsto x^{\beta_i}$. (Thus M is a finitely-generated A-module.) If $\gamma \in G$, define a simplicial complex Δ_γ on the vertex set V having faces $\{\beta_{i_1}, \ldots, \beta_{i_r}\}$ such that $\gamma - \beta_{i_1} - \cdots - \beta_{i_r} \in G$. Then the Betti numbers $\beta_i^A(M) := \dim_k \operatorname{Tor}_i^A(M, k)$ satisfy*

$$\beta_i^A(M) = \sum_{\gamma \in G} \dim_k \widetilde{H}_{i-1}(\Delta_\gamma; k) \,.$$

8   Reciprocity

Some reciprocity theorems in the theory of linear diophantine equations which were originally proved by combinatorial methods find a pleasing explanation in the setting of Cohen–Macaulay modules and local cohomology. For instance, the formula

$$F(R_\Phi, \lambda)_\infty = (-1)^d \sum_{\substack{\beta \in \overline{E}_\Phi \\ \operatorname{supp}_- \beta = \operatorname{supp} \mathcal{P}}} \lambda^\beta$$

first proved in 1973 [147, Thm. 4.1] now follows from the reciprocity of local cohomology (Theorem 6.4), the Cohen–Macaulayness of R_Φ (Corollary

7.6), and Corollary 7.5. In the same way we could derive the following formula, implicit in [148, Thm. 10.2]: If there exists a $\gamma = (\gamma_1, \ldots, \gamma_n) \in \mathbb{Q}^n$, $-1 < \gamma_i \leq 0$, such that $\Phi\gamma = \alpha$, then

$$F\left(M_{\Phi,\alpha}, \lambda\right)_{\infty} = (-1)^d \sum_{\substack{\beta \in \overline{E}_{\Phi,\alpha} \\ \text{supp}_- \beta = \text{supp}\, \mathcal{P}}} \lambda^\beta . \tag{3}$$

In fact, it is clear from the previous section that the following more general statement is valid.

8.1 Theorem. *If $M_{\Phi,\alpha}$ is Cohen–Macaulay then (3) holds.*

Leaving the Cohen–Macaualy case the formula (3) will in general fail. However, using the local cohomology expansion one can still get an exact formula revealing the error.

8.2. Reciprocity Theorem. *Let $d = \dim M_{\Phi,\alpha}$ and $e = \operatorname{depth} M_{\Phi,\alpha}$. Then*

$$\begin{aligned}
F\left(M_{\Phi,\alpha}, \lambda\right)_{\infty} &= \sum_{i=e}^{d} (-1)^i \left[\sum_{\beta \in \overline{E}_{\Phi,\alpha}} (\dim_k \widetilde{H}_{d-i-1}(\Gamma_\beta; k)) \lambda^\beta \right] \\
&= \sum_{\beta \in \overline{E}_{\Phi,\alpha}} \tilde{\chi}(\Gamma_\beta) \lambda^\beta ,
\end{aligned}$$

where $\tilde{\chi}$ denotes reduced Euler characteristic.

Proof. Insert the formula for $F(H^i(M_{\Phi,\alpha}), \lambda)$ obtained from Theorem 7.3 into the reciprocity formula of local cohomology (Theorem 6.4). $\square$

The main term ($i = d$) corresponds to the right hand side of (3) and the other terms are corrections. It is now evident when these corrections vanish. In particular, Cohen–Macaulayness is not necessary for (3).

8.3 Corollary. *(3) holds if and only if for all $\beta \in \overline{E}_{\Phi,\alpha}$ such that $\Gamma_\beta \neq \emptyset$, we have $\tilde{\chi}(\Gamma_\beta) = 0$.*

Let $P(n) \in \mathbb{C}[n]$, $\deg P = d - 1$, and let

$$F(x) = \sum_{n \geq 0} P(n)x^n = W(x)(1 - x)^{-d} , \quad \deg W < d .$$

The following reciprocity theorem is actually true not just for polynomials $P(n)$, but for functions $\sum_{i=1}^{s} P_i(n)\gamma_i^n$, where $P_i(n) \in \mathbb{C}[n]$ and $0 \neq \gamma_i \in \mathbb{C}$. However, we will only need the polynomial case. Even more general results were given in [137, §44] and [140, §3], with an explicit statement of Theorem 8.4 itself for functions $\sum P_i(n)\gamma_i^n$ appearing in [138, p. 5].

8.4 Theorem. $F(x)_\infty = -\sum_{n<0} P(n)x^n$.

Proof. Consider the $\mathbb{C}[x]$-module $\{\sum_{n\in\mathbb{Z}} f(n)x^n \mid f : \mathbb{Z} \to \mathbb{C}\}$. Let $G(x) = \sum_{n\in\mathbb{Z}} P(n)x^n$. Since the dth difference is 0 for a polynomial of degree $d-1$, i.e., $\sum_{i=0}^{d}(-1)^{d-i}\binom{d}{i}P(n+i) = 0$ for all n, we get $(1-x)^d G(x) = 0$. Hence, $W(x) = (1-x)^d F(x) = -(1-x)^d \sum_{n<0} P(n)x^n$ which shows that $F(x)$ and $-\sum_{n<0} P(n)x^n$ are equal as rational functions. $\qquad\square$

9   Reciprocity for integer stochastic matrices

Let us once more return to the problem of enumerating magic squares. Recall that $H_n(r)$ denotes the polynomial which for $r \geq 0$ counts the number of $n \times n$ $\mathbb{N}$-matrices having line sum r for all lines (i.e., rows and columns). Also, for $r < 0$ let $\overline{H}_n(r)$ be the number of $n \times n$ matrices of strictly negative integers having constant line sum r. Set $F(x) := \sum_{r\geq 0} H_n(r)x^r$. By our earlier reciprocity results

$$F(x)_\infty = (-1)^{(n-1)^2+1} \sum_{r<0} \overline{H}_n(r)x^r \ ,$$

and by Theorem 8.4,

$$F(x)_\infty = -\sum_{r<0} H_n(r)x^r \ .$$

Hence, $\overline{H}_n(r) = (-1)^{n-1}H_n(-r)$ for all $r < 0$. There exists a simple transformation between the positive and strictly negative cases as follows: M is an $\mathbb{N}$-matrix with line sum r if and only if $-M - J$ is a matrix of strictly negative integers having line sum $-n - r$, where J is the $n \times n$ matrix of all 1's. This bijection shows that $H_n(r) = \overline{H}_n(-n - r)$. Hence, we conclude that

$$H_n(r) = (-1)^{n-1}H_n(-n - r) , \quad \text{and}$$
$$H_n(-1) = H_n(-2) = \cdots = H_n(-n + 1) = 0 \ .$$

All parts (i)–(iv) of the Anand–Dumir–Gupta conjecture have now been verified. We remark that with this information it is possible to explicitly determine the polynomials $H_n(r)$ for small values of n. For instance, for the case $n = 3$ (first done by MacMahon) we know that $H_3(-1) = H_3(-2) = 0$, $H_3(0) = H_3(-3) = 1$ and $H_3(1) = H_3(-4) = 6$, and being a polynomial of degree $(3-1)^2 = 4$ these six values determine $H_3(r)$. With the aid of a computer $H_n(r)$ has been explicitly computed up to $n = 6$ [102].

10 Rational points in integral polytopes

A topic closely related to the theory of linear diophantine equations is that of integral convex polytopes, which we will now mention in passing. Let $\mathcal{P} \subset \mathbb{R}^n$ be a convex d-dimensional polytope with vertices $\mathbb{Z}^n$. For $m > 0$ let

$$
\begin{aligned}
i(\mathcal{P}, m) &= \#\{\alpha \in \mathcal{P} \mid m\alpha \in \mathbb{Z}^n\}, \\
\bar{i}(\mathcal{P}, m) &= \#\{\alpha \in \mathcal{P} - \partial\mathcal{P} \mid m\alpha \in \mathbb{Z}^n\}.
\end{aligned}
$$

For instance, if $\mathcal{P}$ is the square in $\mathbb{R}^2$ having vertices $(0,0), (1,0), (0,1)$ and $(1,1)$ then $i(\mathcal{P}, m) = (m+1)^2$ and $\bar{i}(\mathcal{P}, m) = (m-1)^2$. The following result is due to Ehrhart [69] [70] (made more precise by Macdonald [117] [118]). It also follows from [147], was proved independently in [124] and [158], and is a simple consequence of Corollary 3.7, Corollary 7.6(b), Theorem 8.1, and Theorem 8.4.

10.1 Theorem. *$i(\mathcal{P}, m)$ and $\bar{i}(\mathcal{P}, m)$ are polynomials of degree d, $i(\mathcal{P}, 0) = 1$, and $\bar{i}(\mathcal{P}, m) = (-1)^d i(\mathcal{P}, -m)$.*

10.2 Example. Let $\mathcal{P} = \Omega_n = \{\text{doubly stochastic } n \times n \text{ matrices}\} \subset \mathbb{R}^{n^2}$. Ω_n is a convex polytope of dimension $(n-1)^2$, and by the Birkhoff–von Neumann theorem its vertices are the permutation matrices. One finds that $i(\Omega_n, m) = \#\{n \times n \text{ N-matrices with constant line sum } m\} = H_n(m)$ and $\bar{i}(\Omega_n, m) = \#\{n \times n \text{ P-matrices with constant line sum } m\} = \overline{H}_n(-m)$. Thus the earlier results on the enumeration of magic squares (Anand–Dumir–Gupta conjecture) can also be derived via the preceding theorem.

10.3 Theorem. *Let $d = n$. Then $i(\mathcal{P}, m) = (\mathrm{vol}\,\mathcal{P})m^d +$ lower terms.*

Proof. Fix $m > 0$. For every $\alpha \in \mathcal{P}$ such that $m\alpha \in \mathbb{Z}^d$, surround α by a d-dimensional cube of side m^{-1}. There are by definition $i(\mathcal{P}, m)$ such cubes, each of volume m^{-d}. Hence, essentially by the definition of the Riemann integral, we get $\lim_{m \to \infty} i(\mathcal{P}, m)m^{-d} = \mathrm{vol}\,\mathcal{P}$. $\quad\square$

10.4 Corollary. *Any d values of $i(\mathcal{P}, m)$, $m \geq 1$, or $\bar{i}(\mathcal{P}, m)$, $m \geq 1$, determine the volume of $\mathcal{P}$.*

For instance, for $d = 2$ we see that $\mathrm{vol}\,\mathcal{P}$ is determined by $i(\mathcal{P}, 1)$ and $\bar{i}(\mathcal{P}, 1)$. Such a result was proved by Pick [136] for nonconvex polygonal regions, and Pick's theorem can be obtained by subdivision from the convex case considered here. Observe that in terms of $d + 1$ consecutive values of $i(\mathcal{P}, m)$ we have

$$\mathrm{vol}\,\mathcal{P} = \frac{1}{d!} \sum_{j=0}^{d} (-1)^{d-j} \binom{d}{j} i(\mathcal{P}, m + j) \,.$$

11   Free resolutions

For further developments we shall need to review again some algebraic background. Let R be a finitely-generated $\mathbb{N}^m$-graded k-algebra and M a finitely-generated $\mathbb{Z}^m$-graded R-module. As usual M can be regarded as a finitely-generated module over a polynomial ring A. If $x_1, x_2, \ldots, x_s \in \mathcal{H}(R_+)$ generate R, then make the surjection $A = k[y_1, y_2, \ldots, y_s] \to R$, $y_i \mapsto x_i$, degree-preserving by setting $\deg y_i = \deg x_i$.

11.1 Definition. A *finite free resolution* (f.f.r.) of M (as a graded A-module) is an exact sequence

$$0 \to \Lambda_t \xrightarrow{\phi_t} \Lambda_{t-1} \xrightarrow{\phi_{t-1}} \cdots \xrightarrow{\phi_2} \Lambda_1 \xrightarrow{\phi_1} \Lambda_0 \xrightarrow{\phi_0} M \to 0 \quad (4)$$

where the Λ_i's are free finitely-generated graded A-modules:

$$\Lambda_i \cong A(\alpha_{1i}) \oplus A(\alpha_{2i}) \oplus \cdots \oplus A(\alpha_{q_i i}) \,, \quad \alpha_{ji} \in \mathbb{Z}^m \,,$$

and $A(\alpha) \cong A$ with grading $A(\alpha)_\beta \cong A_{\beta - \alpha}$, and where the maps ϕ_i are degree-preserving.

   The *homological dimension* of M, $\mathrm{hd}_A\, M$, is the minimal t possible in (4). By the Hilbert syzygy theorem $\mathrm{hd}_A\, M \leq s = \dim A$. The following sharper result is due to Auslander and Buchsbaum.

11.2 Theorem. $\mathrm{hd}_A\, M = s - \mathrm{depth}\, M$.

   The f.f.r. (4) is said to be *minimal* if each Λ_i has smallest possible rank (it can easily be shown that these ranks can be simultaneously minimized). If (4) is minimal let $\beta_i^A(M) := \mathrm{rank}\, \Lambda_i$. An equivalent definition of the *Betti numbers* β_i is $\beta_i^A(M) = \dim_k \mathrm{Tor}_i^A(M, k)$.

   The Hilbert series can be read off immediately from any finite free resolution (4).

11.3 Theorem. $F(M, \lambda) = \sum_{i=0}^{t} (-1)^i F(\Lambda_i, \lambda)$

$$= \frac{\sum_{i=0}^{t} (-1)^i \left(\lambda^{\alpha_{1i}} + \cdots + \lambda^{\alpha_{q_i i}} \right)}{\prod_{j=1}^{s} \left(1 - \lambda^{\deg y_j} \right)} .$$

11.4 Example. Let $A = k[x, y, z, w]$, $R = A/(xyz, xw, yw, zw)$, and for simplicity let $\deg x = \deg y = \deg z = \deg w = 1$. Consider the following f.f.r. of R:

$$0 \to A \xrightarrow{\phi_3} A^4 \xrightarrow{\phi_2} A^4 \xrightarrow{\phi_1} A \to R \to 0$$

$$\begin{bmatrix} z & -y & x & 0 \end{bmatrix} \quad \begin{bmatrix} 0 & y & -x & 0 \\ 0 & z & 0 & -x \\ 0 & 0 & z & -y \\ w & -yz & 0 & 0 \end{bmatrix} \quad \begin{bmatrix} xyz \\ xw \\ yz \\ zw \end{bmatrix}$$

This f.f.r. can quickly be seen to be minimal (assuming it is indeed an f.f.r.) using the following criterion.

11.5 Proposition. *An f.f.r. is minimal if and only if no matrix entry belongs to k^*.*

The degrees of basic elements in the above f.f.r. are

$$0 \longrightarrow A \longrightarrow A^4 \longrightarrow A^4 \longrightarrow A \longrightarrow R \longrightarrow 0$$
$$\quad\quad 4 \quad\quad 3,3,3,4 \quad 3,2,2,2 \quad\quad 0$$

so we can read off the Hilbert series

$$F(R, \lambda) = \frac{1 - (3\lambda^2 + \lambda^3) + (3\lambda^3 + \lambda^4) - \lambda^4}{(1 - \lambda)^4} .$$

Now $\mathrm{hd}_A R = 3 = 4 - \mathrm{depth}\, R$, so depth $R = 1$. But $\dim R = 2$, so R is not Cohen–Macaulay.

If a Hilbert series has been computed by other means it is not in general possible to decompose as to see the Betti numbers. For instance, with A and R as in the previous example let $R_1 = A/(xz, xw, yw)$. Then $F(R_1, \lambda) = F(R, \lambda) = (1-\lambda)^{-4}(1-3\lambda^2+2\lambda^3)$, but R_1 is Cohen–Macaulay and R is not. From the numerator $1-3\lambda^2+2\lambda^3$ of $F(R, \lambda)$ it is impossible to determine (without further information) the "correct" decomposition $1 - (3\lambda^2 + \lambda^3) + (3\lambda^3 + \lambda^4) - \lambda^4$.

12   Duality and canonical modules

An f.f.r. (4) can be dualized by applying the functor $\text{Hom}_A(\cdot, A)$. If $\Lambda = A(\alpha_1)\oplus\cdots\oplus A(\alpha_q)$ then $\Lambda^* = \text{Hom}_A(\Lambda, A) = A(-\alpha_1)\oplus\cdots\oplus A(-\alpha_q)$, so Λ^* is a free module of the same rank but with a different grading. All arrows are reversed, and the matrices expressing these arrows change to their transposes. For instance, the dual of the f.f.r. of the preceding example is as follows:

$$0 \xrightarrow{\phi_0^*} A \xrightarrow{\phi_1^*} A^4 \xrightarrow{\phi_2^*} A^4 \xrightarrow{\phi_3^*} A \xrightarrow{\phi_4^*} 0$$

$$[xyz \quad xw \quad yw \quad zw] \quad \begin{bmatrix} 0 & 0 & 0 & w \\ y & z & 0 & -yz \\ -x & 0 & z & 0 \\ 0 & -x & -y & 0 \end{bmatrix} \quad \begin{bmatrix} z \\ -y \\ x \\ 0 \end{bmatrix}.$$

Clearly $\phi_{i+1}^*\phi_i^* = 0$, so the dualized resolution is a complex, but it is in general not exact. The homology of a dualized (minimal) free resolution of an R-module M, considered as A-module, is one of the fundamental functors of homological algebra (which is independent of the free resolution, minimal or not).

12.1 Definition. $\text{Ext}_A^i(M, A) = \ker \phi_{i+1}^* / \text{im } \phi_i^*$.   The *injective hull* of k as an $A = k[y_1, \ldots, y_s]$-module is $E_A(k) = k[y_1^{-1}, \ldots, y_s^{-1}]$. This given we define the (Matlis) *dual module* of any $\mathbb{Z}^m$-graded A-module M by

$$M^\vee = \text{Hom}_A(M, E_A(k)) .$$

$M^\vee$ is made into a graded module by saying that $\phi : M \to E_A(k)$ has degree α if $\phi(M_\beta) \subseteq E_A(k)_{\beta-\alpha}$ for all $\beta \in \mathbb{Z}^m$. The module $M^\vee$ is always artinian (i.e., satisfies the descending chain condition on submodules), but will not be finitely-generated unless M is artinian (which is the same as $\dim_k M < \infty$ since we are assuming M is finitely-generated).

Fact 1. $F(M^\vee, \lambda) = F(M, \lambda^{-1})$.

Fact 2. $M^{\vee\vee} = \widehat{M}$, the A_+-adic completion of M.

12.2 Example. Let $A = k[x, y]$, $R = A/(xy)$, $\deg x = (1, 0)$ and $\deg y = (0, 1)$. A homogeneous k-basis for $R^\vee$ consists of those $\phi : R \to k[x^{-1}, y^{-1}]$ such that $\phi(1) = 1$ or $\phi(1) = x^{-n}, n > 0$, or $\phi(1) = y^{-n}, n > 0$, since we cannot have negative exponents of both x and y appear in the image of the element 1. Thus,

$$F(R^\vee, \lambda) \;=\; 1 + \sum_{n>0} \left(\lambda_1^{-n} + \lambda_2^{-n} \right) , \quad \text{and}$$

$$F(R, \lambda) \;=\; 1 + \sum_{n>0} \left(\lambda_1^n + \lambda_2^n \right) .$$

The functors Ext and $^\vee$ are related to local cohomology by the following remarkable result [87, §6] [48, Ch. 3.5 and Thm. 3.6.19].

12.3 Local Duality Theorem. $\mathrm{Ext}_A^i(M, A)^\vee = H^{s-i}(M)$.

Let M be a Cohen–Macaulay module of dimension d with a minimal free resolution

$$0 \longrightarrow \Lambda_t \xrightarrow{\phi_t} \Lambda_{t-1} \xrightarrow{\phi_{t-1}} \cdots \xrightarrow{\phi_1} \Lambda_0 \longrightarrow M \longrightarrow 0 .$$

Let $\Omega(M) = \mathrm{coker}\, \phi_t^* = \Lambda_t^* / \mathrm{im}\, \phi_t^* = \mathrm{Ext}_A^{s-d}(M, A)$. Equivalently, $\Omega(M)$ is the unique finitely-generated R-module whose completion $\hat{\Omega}(M) = \Omega(M) \otimes_R \hat{R}$ is isomorphic to $H^d(M^\vee)$. Then

$$0 \longrightarrow \Lambda_0^* \xrightarrow{\phi_1^*} \Lambda_1^* \longrightarrow \cdots \xrightarrow{\phi_t^*} \Lambda_t^* \longrightarrow \Omega(M) \to 0 \qquad (5)$$

is an exact sequence, because Cohen–Macaulayness ensures that $H^i(M) \neq 0$ only for $i = d$, hence by local duality $\mathrm{Ext}_A^i(M, A) \neq 0$ only for $i = s - d = \mathrm{hd}_A M = t$. In fact, (5) is a minimal free resolution of $\Omega(M)$. $\Omega(M)$ is called the *canonical module* of M, and it can be shown directly that as an R-module $\Omega(M)$ is independent of A. It is seen from (4) that the Betti numbers of $\Omega(M)$ are the reverse of those of M:

$$\beta_i^A(\Omega(M)) = \beta_{t-i}^A(M) , \quad t = \mathrm{hd}_A M .$$

$\Omega(M)$ has a natural $\mathbb{Z}^m$-grading such that $F(\Omega(M), \lambda) = (-1)^d F(M, 1/\lambda)$ as rational functions. In the following table we record the way that the Hilbert series varies with the fundamental modules associated with the Cohen–Macaulay module M. The subscripts 0 and ∞ signify expansion of a rational function around the origin and infinity respectively.

Module	Hilbert series	
M	$F(M, \lambda)_0$	$= \sum_\alpha h_\alpha \lambda^\alpha$
$M^\vee$	$F(M, 1/\lambda)_\infty$	$= \sum_\alpha h_\alpha \lambda^{-\alpha}$
$\Omega(M)$	$(-1)^d F(M, 1/\lambda)_0$	$= \sum_\alpha \bar{h}_\alpha \lambda^\alpha$
$\Omega(M)^\vee = H^d(M)$	$(-1)^d F(M, \lambda)_\infty$	$= \sum_\alpha \bar{h}_\alpha \lambda^{-\alpha}$

Define the *socle* of a module M by $\operatorname{soc} M := \{u \in M \mid R_+ u = 0\}$. It follows from noetherianness that $\dim_k \operatorname{soc} M < \infty$.

12.4 Theorem. *Let M be a Cohen–Macaulay module of dimension d over $A = k[x_1, \ldots, x_s]$. Then the following numbers are equal:*

(a) $\beta^A_{s-d}(M)$

(b) *the minimum number of generators of $\Omega(M)$ (as an A-module or an R-module)*

(c) $\dim_k \operatorname{soc} H^d(M)$

(d) $\dim_k \operatorname{soc} M/(\theta_1 M + \cdots + \theta_d M)$ *for any h.s.o.p. $\theta_1, \ldots, \theta_d$.*

Proof. The equivalence of (a), (b) and (c) follows from properties mentioned earlier, such as $\hat{\Omega}(M) = H^d(M)^\vee$, etc. (c) equals (d) by straightforward use of the long exact sequence for local cohomology. □

The number just characterized is called the *type* of M.

12.5 Theorem. *Let $R = A/I$ be Cohen–Macaulay. Then the following are equivalent:*

(a) *type $R = 1$,*

(b) $\Omega(R) \cong R$ *(up to a shift in grading).*

A Cohen–Macaulay ring of type one is said to be *Gorenstein*. Thus a minimal free resolution of a Gorenstein ring is "self-dual." In particular,

$$\beta^A_i(R) = \beta^A_{s-d-i}(R) .$$

12.6 Theorem. *If R is Gorenstein then for some $\alpha \in \mathbb{Z}^m$,*

$$F(R, 1/\lambda) = (-1)^d \lambda^\alpha F(R, \lambda) .$$

Proof. $F(R, 1/\lambda)_0 = (-1)^d F(\Omega(R), \lambda) = (-1)^d \lambda^\alpha F(R, \lambda).$ □

If $m = 1$ and R is Gorenstein with Hilbert series

$$F(R, \lambda) = \frac{h_0 + h_1\lambda + \cdots + h_t\lambda^t}{\prod_{i=1}^{d}(1 - \lambda^{\gamma_i})}, \qquad h_t \neq 0,$$

then by the previous theorem $h_i = h_{t-i}$, $i = 0, 1, \ldots, t$, and $\alpha = t - \Sigma\gamma_i = \deg F = \max\{j \mid H^d(R)_j \neq 0\}$. Also, if $\alpha \geq 0$ and each $\gamma_i = 1$ then α is the last value where the Hilbert function and the Hilbert polynomial disagree.

The converse to the preceding theorem is false. For instance, the ring $k[x, y]/(x^3, xy, y^2)$, $\deg x = \deg y = 1$, is Cohen–Macaulay, artinian and $F(R, \lambda) = \lambda^2 F(R, 1/\lambda)$, but is not Gorenstein. For a reduced counterexample one can take $k[x, y, z, w]/(xyz, xw, yw)$. In the positive direction the following can be said [155, Thm. 4.4].

12.7 Theorem. *If R is an $\mathbb{N}^m$-graded Cohen–Macaulay domain, then R is Gorenstein if and only if $F(R, \lambda) = (-1)^d\lambda^\alpha F(R, 1/\lambda)$ for some $\alpha \in \mathbb{Z}^m$.*

Let us now review a few more facts about canonical modules of Cohen–Macaulay rings. Two references are [88] and [48, Ch. 3].

12.8 Theorem. *$\Omega(R)$ is isomorphic to an ideal I of R if and only if $R_\wp$ is Gorenstein for every minimal prime $\wp$ (e.g., if R is a domain).*

If $m = 1$ we can obtain an isomorphism $\Omega(R) \cong I$ as *graded* modules, up to a shift in grading. This is in general false for $m > 1$. Take $R = k[x, y, z]/(xy, xz, yz)$, $\deg x = (1, 0, 0), \deg y = (0, 1, 0)$ and $\deg z = (0, 0, 1)$. Then R is a Cohen–Macaulay ring, the localization at every minimal prime is a field, and $\Omega(R) \cong (x - y, x - z)$, but there is no way of realizing $\Omega(R)$ as a homogeneous ideal. However, if $m > 1$ and R is a domain one can realize $\Omega(R) \cong I$ as graded modules up to a shift in grading.

12.9 Theorem. *If $\Omega(R) \cong I$ then R/I is Gorenstein and either $I = R$ or $\dim R/I = \dim R - 1$.*

12.10 Theorem. *Let $\theta_1, \ldots, \theta_d$ be an h.s.o.p. for M and $S = k[\theta_1, \ldots, \theta_d]$. Then $\Omega(M) \cong \mathrm{Hom}_S(M, S)$.*

The isomorphism here is as R-modules. There is a standard way of making $\mathrm{Hom}_S(M, S)$ into an R-module: if $x \in R$, $\phi \in \mathrm{Hom}_S(M, S)$ and $u \in M$, define $(x\phi)(u) = \phi(xu)$.

13   A final look at linear equations

We shall now return for the last time to the rings R_Φ of linear diophantine equations. Recall that Φ is an $r \times n$ $\mathbb{Z}$-matrix of maximal rank, $E_\Phi = \{\beta \in \mathbb{N}^n \,|\, \Phi\beta = 0\}$ and $R_\Phi = kE_\Phi$, the monoid algebra of E_Φ over k. The following discussion could be extended to the modules $M_{\Phi,\alpha}$, but for simplicity we consider only R_Φ which is always Cohen–Macaulay. Assume there exists $\beta \in E_\Phi$ such that $\beta > 0$. Recall that $H^d(R_\Phi) \cong k\{x^\beta \,|\, \Phi\beta = 0, \beta < 0\}$ and that in general $\hat{\Omega}(M) = H^d(M)^\vee$.

13.1 Corollary. $\Omega(R_\Phi) \cong k\{x^\beta \,|\, \beta \in E_\Phi,\ \beta > 0\}$.

Thus, $\Omega(R_\Phi)$ is isomorphic to an ideal in R_Φ. Since R_Φ is a domain we know $\Omega(R_\Phi)$ can in fact be realized as a graded ideal, and the above corollary identifies this ideal.

13.2 Corollary. R_Φ is Gorenstein if and only if there exists a unique minimal $\beta > 0$ in E_Φ (i.e., if $\gamma > 0$, $\gamma \in E_\Phi$, then $\gamma - \beta \geq 0$).

13.3 Corollary. R_Φ is Gorenstein if $(1, 1, \ldots, 1) \in E_\Phi$.

The last result has a nice equivalent formulation in terms of invariant theory: if $T \subseteq SL_n(k)$ is a torus acting on $R = k[x_1, \ldots, x_n]$, then R^T is Gorenstein. In this connection we would like to mention the following conjecture of Hochster, Stanley and others: If $G \subseteq SL_n(k)$ is linearly reductive, then R^G is Gorenstein. This is known to be true for finite groups (Watanabe [183]), tori (just shown) and semisimple groups (Hochster and Roberts [96]). However, a counterexample was given by Knop [112] (see also [48, Exer. 6.5.8]). Also, R^G is known to be Cohen–Macaulay for any linearly reductive $G \subseteq GL_n(k)$ (Hochster and Roberts [96]). For further information see [48, §6.5].

Finally, consider again the algebra of magic squares. Let E_Φ be the set of $n \times n$ $\mathbb{N}$-matrices having equal line sums.

$$\begin{bmatrix} 1 & \cdots & 1 \\ \vdots & & \vdots \\ 1 & \cdots & 1 \end{bmatrix} \in E_\Phi, \qquad \text{hence } R_\Phi \text{ is Gorenstein,}$$

$$\text{hence } H_n(R) = (-1)^{n-1} H_n(-n - r).$$

Conversely, if the last equality is proved combinatorially, which can be done, then $F(R_\Phi, \lambda) = (-1)^d \lambda^\alpha F(R_\Phi, 1/\lambda)$, which by Theorem 12.7 implies that R_Φ is Gorenstein since R_Φ is a domain. The same arguments go through also for symmetric magic squares.

Chapter II

The Face Ring of a Simplicial Complex

1 Elementary properties of the face ring

Let Δ be a finite simplicial complex on the vertex set $V = \{x_1, \ldots, x_n\}$. Recall that this means that Δ is a collection of subsets of V such that $F \subseteq G \in \Delta \Rightarrow F \in \Delta$ and $\{x_i\} \in \Delta$ for all $x_i \in V$. The elements of Δ are called *faces*. If $F \in \Delta$, then define $\dim F := |F| - 1$ and $\dim \Delta := \max_{F \in \Delta}(\dim F)$. Let $d = \dim \Delta + 1$. Given any field k we now define the *face ring* (or *Stanley-Reisner ring*) $k[\Delta]$ of the complex Δ.

1.1 Definition. $k[\Delta] = k[x_1, \ldots, x_n]/I_\Delta$, where

$$I_\Delta = (x_{i_1} x_{i_2} \cdots x_{i_r} \mid i_1 < i_2 < \cdots < i_r \,, \quad \{x_{i_1}, x_{i_2}, \ldots, x_{i_r}\} \notin \Delta) \ .$$

1.2 Example. Consider the following plane projection of a triangulation of the 2-sphere

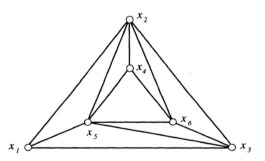

Here $I_\Delta = (x_1 x_4, x_1 x_6, x_3 x_4, x_2 x_5 x_6, \ x_2 x_3 x_5)$.

1.3 Theorem. $\dim k[\Delta] = 1 + \dim \Delta = d$.

Proof. $\dim k[\Delta]$ = maximal cardinality of an algebraically independent set of vertices $x_{i_1}, \ldots, x_{i_j}$ = maximal cardinality of any face.   □

Let f_i be the number of i-dimensional faces of Δ. Since $\emptyset \in \Delta$ (unless $\Delta = \emptyset$) and $\dim \emptyset = -1$, we get $f_{-1} = 1$ for $\Delta \neq \emptyset$. Also, $f_0 = |V|$. The d-tuple $f(\Delta) = (f_0, f_1, \ldots, f_{d-1})$ is called the *f-vector* of Δ. The theme for much of the following is to glean combinatorial information on f-vectors from algebraic information on face rings.

1.4 Theorem. *Define* $\deg x_i = 1$. *Then*

$$H(k[\Delta], m) = \begin{cases} 1, & m = 0 \\ \sum_{i=0}^{d-1} f_i \binom{m-1}{i}, & m > 0. \end{cases}$$

Equivalently,

$$F(k[\Delta], \lambda) = \sum_{i=-1}^{d-1} \frac{f_i \lambda^{i+1}}{(1-\lambda)^{i+1}}.$$

Note that the expression $\sum f_i \binom{m-1}{i}$ evaluated at $m = 0$ gives the Euler characteristic of Δ. Thus, the Hilbert function of $k[\Delta]$ lacks exceptional values if and only if $\chi(\Delta) = 1$. To prove the above theorem it is easiest to work with a finer grading and then specialize. Define the *fine grading* of $k[\Delta]$ by $\deg x_i = (0, \ldots, 0, 1, 0, \ldots, 0) \in \mathbb{Z}^n$, the ith unit coordinate vector. Let $\operatorname{supp} x_1^{a_1} x_2^{a_2} \cdots x_n^{a_n} = \{x_i \mid a_i > 0\}$. Clearly, all monomials $u = x_1^{a_1} x_2^{a_2} \cdots x_n^{a_n}$ such that $\operatorname{supp} u \in \Delta$ form a k-basis for $k[\Delta]$. By counting such monomials u according to their support $F \in \Delta$ we arrive at the following expression for the Hilbert series of the fine grading:

$$F(k[\Delta], \lambda) = \sum_{F \in \Delta} \prod_{x_i \in F} \frac{\lambda_i}{1 - \lambda_i}.$$

Now replace all λ_i by λ to obtain Theorem 1.4.

2   *f*-vectors and *h*-vectors of complexes and multicomplexes

What can be said in general about f-vectors of simplicial complexes? There is the following characterization given independently by Schützenberger, Kruskal and Katona in response to a conjecture by Schützenberger (see [85] for references). This result is known as the *Kruskal-Katona theorem*, since it was not realized at first that Schützenberger [143] actually had the first proof. Given two integers, $\ell, i > 0$ write

$$\ell = \binom{n_i}{i} + \binom{n_{i-1}}{i-1} + \cdots + \binom{n_j}{j} , \quad n_i > n_{i-1} > \cdots > n_j \geq j \geq 1 .$$

A unique such expansion exists. Define

$$\ell^{(i)} = \binom{n_i}{i+1} + \binom{n_{i-1}}{i} + \cdots + \binom{n_j}{j+1} .$$

2.1 Theorem (Schützenberger, Kruskal, Katona). *A vector* $(f_0, f_1, \ldots, f_{d-1}) \in \mathbb{Z}^d$ *is the f-vector of some* $(d-1)$-*dimensional simplicial complex* Δ *if and only if*

$$0 < f_{i+1} \leq f_i^{(i+1)} , \quad 0 \leq i \leq d - 2 .$$

For instance, for $i = 0$: $f_1 \leq f_0^{(1)} = \binom{f_0}{2}$. Theorem 2.1 is proved using the following construction. List all i-element subsets $a_1 < a_2 < \cdots < a_i$ of $\mathbb{N}$ in reverse lexicographic order. For instance, for $i = 3$ the list starts 012, 013, 023, 123, 014, 024, 124, 034, 134, 234, 015, 025, ... Given $f = (f_0, f_1, \ldots, f_{d-1})$, $f_i > 0$, let $\Delta_f = \{\emptyset\} \cup \bigcup_{i=0}^{d-1} \{\text{first } f_i \ (i+1)\text{-element sets in above order}\}$. One then verifies that the following are equivalent:

(i) f is the f-vector of a simplicial complex Δ,

(ii) Δ_f is a simplicial complex,

(iii) $f_{i+1} \leq f_i^{(i+1)}$, $i = 0, 1, \ldots, d - 2$,

The difficult implication is (i) $\Rightarrow$ (ii). For a nice proof of Theorem 2.1, see Greene–Kleitman [85, Sect. 8].

Along with simplicial complexes we shall need the more general notion of multicomplexes. A *multicomplex* Γ on $V = \{x_1, \ldots, x_n\}$ is a set of monomials $x_1^{a_1} \cdots x_n^{a_n}$ such that $u \in \Gamma$, $v|u$ implies $v \in \Gamma$. So a simplicial complex corresponds to the case of squarefree monomials. Multicomplexes are sometimes called "semisimplicial complexes" by topologists. For a multicomplex Γ, let $h_i := \#\{u \in \Gamma \mid \deg u = i\}$, and define the *h-vector* $h(\Gamma) = (h_0, h_1, \ldots)$. An h-vector may be infinite, and if $\Gamma \neq \emptyset$

then $h_0 = 1$. If $h_i = 0$ for $i > d$ we also write $h(\Gamma) = (h_0, \ldots, h_d)$. A sequence $(h_0, h_1, \ldots)$ which is the h-vector of some nonempty multicomplex Γ will be called an M-vector (after F. S. Macaulay, because of Theorems 2.2 and 2.3).

Recall the definition of $\ell^{(i)}$, and define in analogy with the earlier notation

$$\ell^{\langle i \rangle} = \binom{n_i + 1}{i + 1} + \binom{n_{i-1} + 1}{i} + \cdots + \binom{n_j + 1}{j + 1}, \quad 0^{\langle i \rangle} = 0 .$$

2.2 Theorem (essentially Macaulay [113]). $(h_0, h_1, \ldots)$ *is an M-vector if and only if* $h_0 = 1$ *and* $0 \leq h_{i+1} \leq h_i^{\langle i \rangle}, i \geq 1$.

Just as in the case of simplicial complexes, list all monomials of degree i in reverse lexicographic order. E.g., for $i = 3$:

$$x_1^3, \; x_1^2 x_2, \; x_1 x_2^2, \; x_2^3, \; x_1^2 x_3, \; x_1 x_2 x_3, \; x_2^2 x_3, \; x_1 x_3^2, \ldots \; .$$

Given $h = (h_0, h_1, \ldots)$ with $h_0 = 1$, let $\Gamma_h = \bigcup_{i \geq 0}$ {first h_i monomials of degree i in above order}. To prove Theorem 2.2, one then verifies that the following are equivalent:

(i) h is an M-vector,

(ii) Γ_h is a multicomplex,

(iii) $0 \leq h_{i+1} \leq h_i^{\langle i \rangle}, i \geq 1$.

Again, the difficult implication is (i) $\Rightarrow$ (ii). Concerning the proof, Macaulay [113, p. 537] states: "The proof of the theorem ... is given only to place it on record. It is too long and complicated to provide any but the most tedious reading." Macaulay's assessment is certainly accurate. Simpler proofs were given by Sperner, Whipple, and finally Clements and Lindström. Clements and Lindström [57] in fact prove a "generalized Macaulay theorem" which contains both the Kruskal-Katona and the Macaulay theorems as special cases.

The preceding enumerative considerations are closely related to the topic of Hilbert functions of graded algebras.

2.3 Theorem (Macaulay [113]). *Let R be an $\mathbb{N}$-graded k-algebra generated by $x_1, \ldots, x_n \in \mathcal{H}(R_+)$. Then R has a k-basis which is a multicomplex on* $\{x_1, \ldots, x_n\}$.

To prove this result one puts the monomials in $x_1, \ldots, x_n$ into reverse lexicographic order, selects a k-basis for R by taking the lexicographically earliest linearly independent subsequence of monomials, and then one shows that this basis forms a multicomplex. For details, see [155, Thm. 2.1].

2.4 Corollary. *Fix a field k. Let $H : \mathbb{N} \to \mathbb{Z}$. Then the following are equivalent:*

(i) $(H(0), H(1), \ldots)$ *is an M-vector,*

(ii) *there exists a graded algebra $R = R_0 \oplus R_1 \oplus \cdots$, generated by R_1, such that $H(R, i) = H(i)$.*

Proof. (ii) $\Rightarrow$ (i) is immediate from the previous theorem.

(i) $\Rightarrow$ (ii): Let Γ be a multicomplex on $\{x_1, \ldots, x_n\}$ such that $h(\Gamma) = (H(0), H(1), \ldots)$, and let $R = k[x_1, \ldots, x_n]/(x_1^{a_1} \ldots x_n^{a_n} \notin \Gamma)$. $\qquad\square$

Now we define the h-vector of a graded algebra. Let $R = R_0 \oplus R_1 \oplus \cdots$ be generated by R_1, $\dim R = d$. Thus,

$$F(R, \lambda) = \frac{h_0 + h_1 \lambda + \cdots}{(1 - \lambda)^d}, \quad h_0 + h_1 \lambda + \cdots \in \mathbb{Z}[\lambda].$$

Call the vector $h(R) = (h_0, h_1, \ldots, h_\ell)$ the *h-vector* of R, where $h_i = 0$ for $i > \ell$. Thus two vectors that differ only in the number of trailing 0's are considered equivalent in this context.

2.5 Corollary. *Fix $d \geq 0$. Let $(h_0, \ldots, h_\ell) \in \mathbb{Z}^{\ell+1}$. The following are equivalent:*

(i) $(h_0, \ldots, h_\ell)$ *is an M-vector*

(ii) *there exists a d-dimensional Cohen–Macaulay graded algebra $R = R_0 \oplus R_1 \oplus \cdots$, generated by R_1, such that $h(R) = (h_0, \ldots, h_\ell)$.*

Proof. (i) $\Rightarrow$ (ii): Let $S = S_0 \oplus \cdots \oplus S_\ell$ be generated by S_1, with $H(S, i) = h_i$ (by the previous theorem). Let $R = S[x_1, \ldots, x_d]$. Then R is Cohen–Macaulay; for a regular sequence of length d we can take $x_1, \ldots, x_d$; and when we mod out by them we are left with S.

(ii) $\Rightarrow$ (i): Let $\theta_1, \ldots, \theta_d$ be an h.s.o.p. for R with $\deg \theta_i = 1$. (Such a choice is always possible if the ground field k is infinite. If k is finite one may need to pass to an infinite extension field, which does not affect the Hilbert function or the Cohen–Macaulay property.) Let $S = R/(\theta_1, \ldots, \theta_d)$. Then $F(R, \lambda) = (1 - \lambda)^{-d} F(S, \lambda)$, so $h_i = H(S, i)$ and $(h_0, \ldots, h_\ell)$ is an M-vector. $\qquad\square$

Having previously defined the h-vectors of multicomplexes and graded rings, we now define the h-*vector* of a simplicial complex Δ as follows:

$$h(\Delta) := h(k[\Delta]) .$$

One observes that $h(\Delta)$ is a finite vector, in fact, $h_i = 0$ for $i > d = \dim \Delta + 1$. This definition is equivalent to the following explicit expression for $h(\Delta) = (h_0, h_1, \ldots, h_d)$ in terms of the f-vector $(f_0, f_1, \ldots, f_{d-1})$ of Δ (letting $f_{-1} = 1$):

$$h_k = \sum_{i=0}^{k}(-1)^{k-i}\binom{d-i}{k-i}f_{i-1} , \quad 0 \le k \le d .$$

For example, if Δ is the boundary of an octahedron then $f(\Delta) = (6, 12, 8)$ and $h(\Delta) = (1, 3, 3, 1)$. Three observations are immediate:

(1) $h_d = (-1)^{d-1}\tilde{\chi}(\Delta)$, where $\tilde{\chi}$ denotes reduced Euler characteristic, i.e., $\tilde{\chi}(\Delta) = \sum_{i \ge -1}(-1)^i f_i = \sum_{i \ge -1}(-1)^i \dim_k \widetilde{H}_i(\Delta; k)$,

(2) $h_0 + h_1 + \cdots + h_d = f_{d-1}$, and

(3) knowing the h-vector of Δ is equivalent to knowing the f-vector of Δ.

3   Cohen–Macaulay complexes and the Upper Bound Conjecture

3.1 Definition. Δ is a *Cohen–Macaulay complex* (more precisely: Cohen–Macaulay *over* k) if the face ring $k[\Delta]$ is Cohen–Macaulay.

3.2 Corollary. *If Δ is Cohen–Macaulay, then $h(\Delta)$ is an M-vector.*

Consider the following simplicial complex Δ:

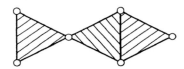

$f(\Delta) = (6, 8, 3)$, so $h(\Delta) = (1, 3, -1, 0)$. This is not an M-vector, so Δ is not Cohen–Macaulay.

A weak converse to Corollary 3.2 is also true, and we get the following result which can be called a "Kruskal–Katona theorem" for Cohen–Macaulay complexes. See [152, Thm. 6].

3.3 Theorem. *Let $h = (h_0, h_1, \ldots, h_d) \in \mathbb{Z}^{d+1}$. Then there exists a $(d-1)$-dimensional Cohen–Macaulay (or shellable, as defined in Chapter 3.2) complex Δ for which $h(\Delta) = h$ if and only if h is an M-vector.*

As a motivation for studying Cohen–Macaulay complexes we want to mention the Upper Bound Conjecture for spheres. This concerns the question: Suppose the geometric realization $|\Delta|$ is homeomorphic to the $(d-1)$-dimensional sphere $\mathbb{S}^{d-1}$. Then given $f_0(\Delta) = n$, how large can f_i be? In order to formulate the conjectured answer we must review the notion of cyclic polytopes.

For $n > d$ let $C(n, d)$ be the convex hull of any n distinct points on the curve $\{(\tau, \tau^2, \ldots, \tau^d) \in \mathbb{R}^d \mid \tau \in \mathbb{R}\}$. The combinatorial type of the *cyclic polytope* $C(n, d)$ is independent of the n points chosen, and $\dim C(n, d) = d$. Cyclic polytopes have been investigated by Carathéodory, Gale, Motzkin, Klee, and others. Here are some known facts.

(a) $C(n, d)$ is simplicial (i.e., every proper face is a simplex), so the boundary $\partial C(n, d)$ defines an abstract simplicial complex $\Delta(n, d)$ such that $|\Delta(n, d)| \cong \mathbb{S}^{d-1}$.

(b) $f_i(\Delta(n, d)) = \binom{n}{i+1}$ for $0 \le i < \left[\frac{d}{2}\right]$.

(c) $f_0, f_1, \ldots, f_{\left[\frac{d}{2}\right]-1}$ determine $f_{\left[\frac{d}{2}\right]}, f_{\left[\frac{d}{2}\right]+1}, \ldots, f_{d-1}$. This is true for any Δ such that $|\Delta| \cong \mathbb{S}^{d-1}$ because of the Dehn–Sommerville equations $h_i = h_{d-i}, 0 \le i \le d$.

3.4 Upper Bound Conjecture (UBC) for Spheres. *If $|\Delta| \cong \mathbb{S}^{d-1}$ and $f_0(\Delta) = n$, then $f_i(\Delta) \le f_i(\Delta(n, d))$, $i = 0, 1, \ldots, d-1$.*

McMullen [122] showed that Δ satisfies UBC if

$$h_i(\Delta) \le \binom{n - d + i - 1}{i} \quad \text{for} \quad 0 \le i < \left[\frac{d}{2}\right].$$

The conjectured upper bound for $f_i(\Delta)$ is true for $i < \left[\frac{d}{2}\right]$ by (b) and plausible for $i \ge \left[\frac{d}{2}\right]$ because of (c). The UBC for simplicial convex polytopes was proposed by Motzkin in 1957 [129] and proved by McMullen [122] in 1970. However, there exist triangulations of spheres, first found by Grünbaum, which are not polytopal. The smallest example of a non-polytopal sphere has parameters $d = 4$ and $f_0 = 8$. Kalai [106] showed that "most" spheres are not polytopal. Klee suggested to extend the UBC to all spheres, and the general result was established by Stanley [150] as follows.

3.5 Corollary. *If $|\Delta| \approx \mathbb{S}^{d-1}$ and Δ is Cohen–Macaulay, then UBC holds for Δ.*

Proof. $h_1(\Delta) = n - d$ and $(h_0, \ldots, h_d)$ is an M-vector. Hence, $h_i \le$ total number of monomials of degree i in $n - d$ variables $= \binom{n-d+i-1}{i}$. □

To complete the proof of the UBC, we need to answer the question, when is a simplicial complex Δ Cohen–Macaulay? A very useful answer in terms of simplicial homology was given by Reisner in his 1974 thesis [139]. More general results were later obtained by Hochster [95]. In particular, the class of Cohen–Macaulay complexes can be seen to include all triangulations of spheres, so the UBC for spheres follows. We will now look at some of these results in more detail.

4 Homological properties of face rings

Let Δ be a simplicial complex, and for $F \in \Delta$ define the *link*

$$\operatorname{lk} F = \{G \in \Delta \mid G \cup F \in \Delta, G \cap F = \emptyset\} .$$

Recall that the fine grading of the face ring $k[\Delta]$ is given by setting $\deg x_i = (0, \ldots, 0, 1, 0, \ldots, 0) \in \mathbb{Z}^n$, the "1" in the ith position.

4.1 Theorem (Hochster, unpublished). *Under the fine grading*

$$F(H^i(k[\Delta]), \lambda) = \sum_{F \in \Delta} \dim_k \widetilde{H}_{i-|F|-1}(\operatorname{lk} F; k) \prod_{x_i \in F} \frac{\lambda_i^{-1}}{1 - \lambda_i^{-1}} .$$

The main goal of this section is to prove this fundamental result.

4.2 Corollary (Reisner [139]). *Δ is Cohen–Macaulay over k if and only if for all $F \in \Delta$ and all $i < \dim(\operatorname{lk} F)$, we have $\widetilde{H}_i(\operatorname{lk} F; k) = 0$.*

Proof. By Hochster's theorem and Theorem 6.3 of Chapter I, $k[\Delta]$ is Cohen–Macaulay if and only if $\widetilde{H}_{i-|F|-1}(\operatorname{lk} F; k) = 0$ for all $i < d$ and all $F \in \Delta$. Assume that Δ is pure, i.e. that all maximal faces of Δ have dimension $d - 1 = \dim \Delta$. Then $\dim(\operatorname{lk} F) = d - |F| - 1$ for all $F \in \Delta$, so $i < d$ if and only if $i - |F| - 1 < \dim(\operatorname{lk} F)$. It only remains to verify that Δ is pure if either of the two conditions hold. Starting from Hochster's condition, if $|F| < d$ then $\widetilde{H}_{-1}(\operatorname{lk} F) = \widetilde{H}_{|F|-|F|-1}(\operatorname{lk} F) = 0$, so lk $F \neq \emptyset$ and F must be contained in a larger face. Starting from Reisner's condition one observes that the same condition holds for all proper links, hence by induction these are pure. If $\dim \Delta \geq 1$ then $\widetilde{H}_0(\Delta) = \widetilde{H}_0(\text{lk } \emptyset) = 0$, so Δ is connected, and this together with the purity of links of vertices shows that Δ is pure. $\qquad\qquad\square$

The following result, due to Munkres [131], shows that the Cohen–Macaulayness of Δ is a topological property. The homology referred to in the second condition is reduced singular and relative singular homology respectively.

4.3 Proposition (Munkres). *Let $X = |\Delta|$. The following are equivalent:*

(i) *for all $F \in \Delta$ and all $i < \dim(\text{lk } F)$, $\widetilde{H}_i(\text{lk } F; k) = 0$,*

(ii) *for all $p \in X$ and all $i < \dim X = d - 1$, $\widetilde{H}_i(X; k)$*
     $= H_i(X, X - p; k) = 0.$

4.4 Corollary. *If $|\Delta| \cong \mathbb{S}^{d-1}$ then Δ is Cohen–Macaulay.*

4.5 Corollary. *The UBC for spheres holds.*

We will now prove Hochster's theorem (Theorem 4.1) on the local cohomology of face rings. Recall the formulation. Let Δ be a finite simplicial complex on vertices $x_1, x_2, \ldots, x_n$, and let k be a field. Give the face ring $R = k[\Delta]$ the fine grading $\deg x_i = (0, \ldots, 0, 1, 0, \ldots, 0) \in \mathbb{Z}^n$. Let $H^i(R)$ be the ith local cohomology module with the induced $\mathbb{Z}^n$-grading. Then

$$F(H^i(R), \lambda) = \sum_{F \in \Delta} \left(\dim_k \widetilde{H}_{i - |F| - 1}(\text{lk } F) \right) \prod_{x_i \in F} \frac{\lambda_i^{-1}}{1 - \lambda_i^{-1}} \, .$$

Proof. The general plan for this proof is similar to what we did for the modules $M_{\Phi, \alpha}$. The idea is due to Hochster (unpublished), and was in fact our inspiration for Theorem 7.3 of the previous chapter.

Let $\mathcal{K}(\mathbf{x}^\infty, R)$ be the complex of Chapter I, Theorem 6.2. Thus

$$
\begin{aligned}
H^i(R) &= H^i(\mathcal{K}(\mathbf{x}^\infty, R)) \\
&= H^i\left(0 \xrightarrow{\delta_0} R \xrightarrow{\delta_1} \coprod_i R_{x_i} \xrightarrow{\delta_2} \coprod_{i<j} R_{x_i x_j} \right. \\
&\qquad\qquad \left. \xrightarrow{\delta_3} \cdots \xrightarrow{\delta_n} R_{x_1 x_2 \cdots x_n} \longrightarrow 0 \right) \\
&= \ker \delta_{i+1} / \operatorname{im} \delta_i \, .
\end{aligned}
$$

Let $F \subseteq \{x_1, x_2, \ldots, x_n\}$. Write R_F for R localized with respect to $\prod_{x_i \in F} x_i$. By inspection then

$$
R_F = \begin{cases} k[\{x_i, x_i^{-1} \mid x_i \in F\} \cup \{x_j \mid x_j \in \operatorname{lk} F\}] \, , & \text{if } F \in \Delta \\ 0 \, , & \text{if } F \notin \Delta \, . \end{cases}
$$

The latter case is clear since if we invert a set whose product is zero then everything disappears. For the former case, notice that if $x_j \notin \operatorname{st} F := \{G \in \Delta \mid G \cup F \in \Delta\}$ then $x_j \prod_{x_i \in F} x_i = 0$ in R. If $F \in \Delta$ we can equivalently write $R_F = k[\{x_i \mid x_i \in \operatorname{st} F\} \cup \{x_i^{-1} \mid x_i \in F\}]$.

We want to compute $\mathcal{K}(\mathbf{x}^\infty, R)_\alpha$ for $\alpha = (\alpha_1, \alpha_2, \ldots, \alpha_n) \in \mathbb{Z}^n$. First, if $\operatorname{supp} \alpha := \{x_i \mid \alpha_i \neq 0\} \notin \Delta$ then $\mathcal{K}(\mathbf{x}^\infty, R)_\alpha$ is zero. So suppose that $\operatorname{supp} \alpha \in \Delta$ and let $F = \{x_i \mid \alpha_i < 0\}$ and $G = \{x_i \mid \alpha_i > 0\}$, $|F| = j$. Let us look at the rth term in $\mathcal{K}(\mathbf{x}^\infty, R)_\alpha$:

$$
\left(\coprod_{i_1 < \cdots < i_r} R_{x_{i_1} \cdots x_{i_r}} \right)_\alpha = \left(\coprod_{\substack{F' \in \Delta \\ |F'| = r}} R_{F'} \right)_\alpha \, .
$$

This is a vector space over k with basis corresponding to all $F' \supseteq F$ such that $|F'| = r$ and $F' \cup G \in \Delta$, i.e, (deleting F) to all $(r-j)$-element faces of $\operatorname{lk}_{\operatorname{st} G} F$ (the link within $\operatorname{st} G$ of F). The maps in $\mathcal{K}(\mathbf{x}^\infty, R)_\alpha$ are Koszul relations. If we fix an orientation of Δ these are coboundary maps except possibly for signs. Thus, by choosing correct signs for the basis elements of each $(R_F)_\alpha$ we may identify $\mathcal{K}(\mathbf{x}^\infty, R)_\alpha$ with the augmented oriented simplicial cochain complex of $\operatorname{lk}_{\operatorname{st} G} F$ with dimension shifted by $j+1$. (Note that in the case of $M_{\Phi,\alpha}$ we got a chain complex, but here a cochain complex.) So,

$$
H^i(R)_\alpha \cong \widetilde{H}^{i-j-1}(\operatorname{lk}_{\operatorname{st} G} F) \cong \widetilde{H}_{i-j-1}(\operatorname{lk}_{\operatorname{st} G} F) \, ,
$$

(since we are working over a field).

4.6 Example. Let Δ be the 1-dimensional complex:

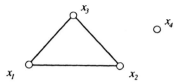

Then $\mathcal{K}(\mathbf{x}^\infty, R)$:

$$0 \to R \to R_{x_1} \oplus R_{x_2} \oplus R_{x_3} \oplus R_{x_4} \to R_{x_1 x_2} \oplus R_{x_1 x_3} \oplus R_{x_2 x_3} \to 0 \, .$$

If $\alpha = (0, 0, 0, 0)$, then $\mathcal{K}(\mathbf{x}^\infty, R)_\alpha$ is the augmented cochain complex for $\mathrm{lk}_{\mathrm{st}\phi}\, \emptyset = \Delta$ and

$$H^i(R)_\alpha = \widetilde{H}_{i-1}(\Delta) \, .$$

If $\alpha = (-2, 3, 0, 0)$, then $0 \to 0 \to (R_{x_1})_\alpha \to (R_{x_1 x_2})_\alpha \to 0$ is the augmented cochain complex for $\mathrm{lk}_{\mathrm{st}\, x_2}\, x_1 = x_2$ and

$$H^i(R)_\alpha = \widetilde{H}_{i-2} \, (\text{point}) \, .$$

Returning to the general proof, it remains to sum over all $\alpha \in \mathbb{Z}^n$. First observe the following simplification. If $G \neq \emptyset$, then $\mathrm{lk}_{\mathrm{st}\, G}\, F$ is a cone over G and therefore acyclic. We may therefore assume that $G = \emptyset$, or equivalently $\alpha \leq 0$, so $\mathrm{lk}_{\mathrm{st}\, G}\, F = \mathrm{lk}_\Delta\, F = \mathrm{lk}\, F$ and $H^i(R)_\alpha \cong \widetilde{H}_{i-|F|-1}(\mathrm{lk}\, F)$. Summing over all α we now get:

$$
\begin{aligned}
F(H^i(R), \lambda) &= \sum_{F \in \Delta} \sum_{\substack{\alpha \leq 0 \\ \mathrm{supp}\, \alpha = F}} \left(\dim_k \widetilde{H}_{i-|F|-1}(\mathrm{lk}\, F) \right) \lambda^\alpha \\
&= \sum_{F \in \Delta} \dim_k \widetilde{H}_{i-|F|-1}(\mathrm{lk}\, F) \prod_{x_i \in F} \frac{\lambda_i^{-1}}{1 - \lambda_i^{-1}} \, ,
\end{aligned}
$$

and the proof is complete. $\qquad\qquad\qquad\qquad\qquad\qquad\qquad\qquad$ $\Box$

Let $\Delta, V = \{x_1, x_2, \ldots, x_n\}$, $A = k[x_1, x_2, \ldots, x_n]$ and $k[\Delta] = A/I_\Delta$ be as before. Consider the $\mathbb{Z}^n$-graded finite-dimensional vector space $T_i = \mathrm{Tor}_i^A(k[\Delta], k)$. We would like to determine its Hilbert series. The answer, given by Hochster, implies a nice formula for the Betti numbers $\beta_i^A(k[\Delta]) = \dim_k \mathrm{Tor}_i^A(k[\Delta], k)$.

4.7 Example. The complex $\Delta : \begin{smallmatrix} \circ\, y \\ | \\ \circ\, x \end{smallmatrix} \;\; \circ\, z$ has a minimal free resolution

$$0 \longrightarrow A \longrightarrow A^2 \longrightarrow A \longrightarrow k[\Delta] \longrightarrow 0$$
$$\begin{array}{cc} [y \ \ -x] & \begin{bmatrix} xz \\ yz \end{bmatrix} \end{array}$$

with degree-preserving maps and generators of degree $(1,1,1)$; $(1,0,1)$ and $(0,1,1)$; and $(0,0,0)$, respectively. Consequently,

$$\begin{aligned} F(T_0, \lambda) &= 1 , \\ F(T_1, \lambda) &= \lambda_1\lambda_3 + \lambda_2\lambda_3 , \\ F(T_2, \lambda) &= \lambda_1\lambda_2\lambda_3 . \end{aligned}$$

4.8 Theorem (Hochster [95]).

$$F(T_i, \lambda) = \sum_{W \subseteq V} \left(\dim_k \widetilde{H}_{|W|-i-1}(\Delta_W) \right) \prod_{x_i \in W} \lambda_i ,$$

where $\Delta_W = \{F \in \Delta \mid F \subseteq W\}$.

The proof is similar to but rather more complicated than the preceding one. Let us merely check that the formula works for the given example:

$$\begin{aligned} F(T_0, \lambda) &= \dim_k \widetilde{H}_{-1}(\emptyset) , \\ F(T_1, \lambda) &= \left(\dim_k \widetilde{H}_0 \begin{pmatrix} x & z \\ \circ & \circ \end{pmatrix} \right) \lambda_1\lambda_3 + \left(\dim_k \widetilde{H}_0 \begin{pmatrix} y & z \\ \circ & \circ \end{pmatrix} \right) \lambda_2\lambda_3 , \\ F(T_2, \lambda) &= \left(\dim_k \widetilde{H}_0 \left(\begin{smallmatrix} \circ\, y \\ | \\ \circ\, x \end{smallmatrix} \; \circ\, z \right) \right) \lambda_1\lambda_2\lambda_3 . \end{aligned}$$

4.9 Corollary. $\beta_i^A(k[\Delta]) = \sum_{W \subseteq V} \dim_k \widetilde{H}_{|W|-i-1}(\Delta_W)$

5　Gorenstein face rings

Let us now turn to this question: which Cohen–Macaulay complexes Δ are *Gorenstein*, i.e., for which Δ is the face ring $k[\Delta]$ a Gorenstein ring? Define for Δ on $V = \{x_1, x_2, \ldots, x_n\}$,

core $V \;\; := \;\; \{x \in V \mid \mathrm{st}\{x\} \neq V\}$, where $\mathrm{st}\{x\} = \{F \in \Delta \mid F \cup \{x\} \in \Delta\}$, and

core $\Delta \;\; := \;\; \Delta_{\mathrm{core}\, V} .$

Thus, $\Delta_{V-\text{core } V}$ is a simplex, and $\Delta = \Delta_{V-\text{core } V} * \text{core } \Delta$ (where $\Delta' * \Delta'' = \{F \cup G | F \in \Delta', G \in \Delta''\}$ is the simplicial join). On the ring-theoretic level:

$$k[\Delta] = k[\text{core } \Delta][x \mid x \in V - \text{core } V].$$

Moreover, Δ and core Δ have the same h-vector (except for 0's at the end).

5.1 Theorem. *Fix a field k (or $\mathbb{Z}$). Let Δ be a simplicial complex and $\Gamma := \text{core } \Delta$. Then the following are equivalent:*

(a) Δ *is Gorenstein (over k),*

(b) *for all $F \in \Gamma$, $\widetilde{H}_i(\text{lk}_\Gamma F; k) = \begin{cases} k, & i = \dim(\text{lk}_\Gamma F) \\ 0, & i < \dim(\text{lk}_\Gamma F) \end{cases},$*

(c) *for $X = |\Gamma|$ and all $p \in X$,*

$$\widetilde{H}_i(X, k) \cong H_i(X, X - p; k) = \begin{cases} k, & i = \dim X \\ 0, & i < \dim X \end{cases},$$

(d) Δ *is Cohen–Macaulay over k, and Γ is an orientable pseudomanifold over k (or even over $\mathbb{Z}$, regardless of k),*

(e) *either* (i) $\Delta = \emptyset, \circ,$ *or* $\circ\ \circ$, *or* (ii) Δ *is Cohen–Macaulay over k of dimension $d - 1 \geq 1$, and the link of every $(d-3)$-face is either a circle or $\circ\!\!-\!\!\circ$ or $\circ\!\!-\!\!\circ\!\!-\!\!\circ$, and $\tilde{\chi}(\Gamma) = (-1)^{\dim \Gamma}$ (the last condition is superfluous over $\mathbb{Z}$ or if char $k = 2$).*

This characterization shows for instance that the 2- and 3-point lines, $\circ\!\!-\!\!\circ$ and $\circ\!\!-\!\!\circ\!\!-\!\!\circ$, are Gorenstein while the 4-point line $\circ\!\!-\!\!\circ\!\!-\!\!\circ\!\!-\!\!\circ$ is not. Condition (e) was given by Hochster [95] (for $k = \mathbb{Z}$) and (b)–(c) by Stanley [152]. Condition (d) stems from a remark by Björner.

Proof. Let as before $T_i = \text{Tor}_i^A(k[\Delta], k)$, $\dim \Delta = d - 1$ and $n = |V|$. Then, $k[\Delta]$ is Gorenstein if and only if $F(T_{n-d+1}, \lambda) = 0$ and $F(T_{n-d}, \lambda) = \prod_{x_i \in W} \lambda_i$ for some $W \subseteq V$. One shows that $W = \text{core } V$ and then uses Hochster's formula for $F(T_{n-d}, \lambda)$ to get (a) $\Leftrightarrow$ (b). The rest is an exercise in combinatorial topology. Let us merely make some remarks concerning condition (d). Suppose that Δ is Cohen–Macaulay over k and $\dim \Gamma = e - 1$. Then Γ is pseudomanifold if and only if every $(e - 2)$-dimensional face of Γ lies in exactly two $(e - 1)$-dimensional faces. Furthermore, Γ is orientable over $k \Leftrightarrow \widetilde{H}_{e-1}(\Gamma; k) \cong k \Leftrightarrow \tilde{\chi}(\Gamma) = (-1)^{e-1}$. In particular, Γ is orientable over k if and only if it is orientable over $\mathbb{Z}$. $\square$

5.2 Corollary. *If $|\Delta| \approx \mathbb{S}^{d-1}$ then Δ is Gorenstein.*

Let R be an $\mathbb{N}^m$-graded finitely-generated Gorenstein algebra. Preserve the usual notation and conventions, in particular $\mathrm{hd}_A R = n - d$. Let $T_i = \mathrm{Tor}_i^A(R, k)$ and $F(T_{n-d}, \lambda) = \lambda^\alpha$. Then,

$$F(T_i, \lambda) = \lambda^\alpha F\left(T_{n-d-i}, 1/\lambda\right) ,$$

where $1/\lambda = (1/\lambda_1, 1/\lambda_2, \ldots, 1/\lambda_m)$. Note that with $\lambda = 1$ this gives the relation $\beta_{n-d-i}^A(R) = \beta_i^A(R)$, which was earlier pointed out.

Now let Δ be a Gorenstein complex such that $\Delta = \mathrm{core}\,\Delta$, and let us apply the preceding formula to $k[\Delta]$. In view of Hochster's formula we get

$$\sum_{W \subseteq V} \left(\dim_k \widetilde{H}_{|W|-i-1}(\Delta_W)\right) \prod_{x_i \in W} \lambda_i$$

$$= \lambda_1 \lambda_2 \cdots \lambda_n \sum_{W \subseteq V} \left(\dim_k \widetilde{H}_{|W|-n+d+i-1}(\Delta_W)\right) \prod_{x_i \in W} \lambda_i^{-1}$$

$$= \sum_{W \subseteq V} \left(\dim_k \widetilde{H}_{|W|-n+d+i-1}(\Delta_W)\right) \prod_{x_i \notin W} \lambda_i .$$

Hence,

$$\widetilde{H}_{|W|-i-1}(\Delta_W) \cong \widetilde{H}_{|V-W|-n+d+i-1}(\Delta_{V-W}) ,$$

i.e., $\widetilde{H}_j(\Delta_W) \cong \widetilde{H}_{d-j-2}(\Delta_{V-W})$ for all $W \subseteq V$ and all j. This is the *Alexander duality theorem* for nonacyclic Gorenstein complexes. (A Gorenstein complex Δ is nonacyclic if and only if $\Delta = \mathrm{core}\,\Delta$, i.e., Δ is not a cone.)

6   Gorenstein Hilbert functions

What can one say about Gorenstein Hilbert functions and about the h-vectors of Gorenstein complexes?

6.1 Definition. A sequence $(h_0, h_1, \ldots, h_s)$, $h_s \neq 0$, is called *Gorenstein* if there exists a Gorenstein algebra $R = R_0 \oplus R_1 \oplus \cdots$ generated by R_1 such that $h(R) = (h_0, h_1, \ldots, h_s)$. (Recall that this means $F(R, \lambda) = (1 - \lambda)^{-d}(h_0 + h_1\lambda + \cdots + h_s\lambda^s)$, where $d = \dim R$.)

Problem 1. Characterize Gorenstein sequences.

Let Δ be a Gorenstein complex with $\dim(\mathrm{core}\,\Delta) = e - 1$. Then the h-vector $h(\mathrm{core}\,\Delta) = (h_0, h_1, \ldots, h_e)$ is a Gorenstein sequence and $h(\Delta) = (h_0, h_1, \ldots, h_e, 0, \ldots, 0)$. In contrast to the Cohen–Macaulay case (Theorem 3.3), not every Gorenstein sequence arises in this way (see fact (a) below).

Problem 2. Characterize the h-vectors $h(\Delta)$ of Gorenstein complexes.

It has earlier been shown that every Gorenstein sequence $(h_0, h_1, \ldots, h_s)$ is symmetric: $h_i = h_{s-i}$. The equations $h_i = h_{s-i}, i = 0, 1, \ldots, \left[\frac{s}{2}\right]$, for the h-vector $h(\Delta) = (h_0, h_1, \ldots, h_s)$ of a complex Δ are known as the *Dehn–Sommerville equations*. Define a *Gorenstein** complex to be a Gorenstein complex Δ for which $\Delta = \text{core } \Delta$, or equivalently, for which Δ is acyclic.

6.2 Conjecture. $(h_0, h_1, \ldots, h_s)$, *where* $h_s \neq 0$, *is the h-vector of some Gorenstein** *complex* Δ *if and only if* $h_i = h_{s-i}$ *for all i and* $(h_0, h_1 - h_0, h_2 - h_1, \ldots, h_{[s/2]} - h_{[s/2]-1})$ *is an M-vector.*

The sufficiency of the conjectured condition follows from a result of Billera and Lee [20], [21]. The necessity is known if Δ is the boundary of a simplicial convex polytope (Stanley [159]) but is otherwise open. See also Chapter 3.1.

Let us review some of the known facts concerning the two above-mentioned problems.

(a) The sequence $(1, 13, 12, 13, 1)$ is Gorenstein (Stanley [155, p. 70]) but not $h(\Delta)$ for some Gorenstein complex Δ (this follows from a result of Klee, cf. [152, p. 58]). For further examples of "bad" Gorenstein sequences, see [15], [40, §3], [41], [42].

(b) Gorenstein sequences and Gorenstein $h(\Delta)$ agree for $h_1 \leq 3$ (follows from a result of Buchsbaum and Eisenbud, cf. [155, Thm. 4.2]).

(c) $(1, 4, 3, 4, 1)$ and $(1, 5, 4, 5, 1)$ are not Gorenstein sequences but are M-vectors (cf. [155, p. 71]). (Improvements to this result were obtained in unpublished work of C. Peskine.)

Proof of (c). Let $R = R_0 \oplus R_1 \oplus \cdots \oplus R_s$ be a 0-dimensional graded Gorenstein algebra generated by R_1, with $R_s \neq 0$. Thus $R_s = \text{soc } R$. If $x \in R_i$, then the principal ideal xR is isomorphic (as an R-module) to R/I for some ideal I. Since $\text{soc } xR = R_s$, there follows the well-known result that R/I is a Gorenstein ring. Hence if $a_j = \dim_k xR_j$, then $(a_0, a_1, \ldots, a_{s-i})$ is a Gorenstein sequence.

Now assume $s = 4$ and $h(R) = (1, 4, 3, 4, 1)$. Let $x \in R_1$ and $xR \cong R/I$. Since R/I is Gorenstein, we have in the above notation $(a_0, a_1, a_2, a_3) = (1, a, a, 1)$ for some $a \geq 1$, so that $h(R/xR) = (1, 3, 3 - a,$

$4 - a)$. But for no $a \geq 1$ is $(1, 3, 3 - a, 4 - a)$ an M-vector, so R does not exist.

Now assume $h(R) = (1, 5, 4, 5, 1)$, and again let $x \in R_1$. Reasoning as above, we have $h(R/xR) = (1, 4, 4 - a, 5 - a)$. In order for this to be an M-vector, we must have $a = 1$. This means in particular that for every $x \in R_1$, $\dim_k xR_1 = 1$. Pick $x, y, z, w \in R_1$ such that xy and zw are linearly independent. (This is possible since $\dim_k R_2 > 1$.) Since $\dim_k xR_1 = 1$, we have $xz = \alpha xy$ for some $\alpha \in k$. Similarly $xz = \beta zw$. Hence $xz = 0$. Similarly $wy = 0$. But since $\dim_k(x + w)R_1 = 1$ we have that $(x + w)y$ and $(x + w)z$ are linearly dependent. But $(x + w)y = xy$ and $(x + w)z = zw$, a contradiction. $\qquad\square$

(d) Let $f(n) = \min\{N \mid (1, n, N, n, 1) \text{ is a Gorenstein sequence}\}$. Thus, $f(4) = 4$, $f(5) = 5$, and $f(13) \leq 12$. We claim:

$$\lim_{n \to \infty} \frac{\log f(n)}{\log n} = \frac{2}{3}.$$

Sketch of Proof. (I am grateful to Peter Kleinschmidt for pointing out a gap in the original unpublished proof of this result.) Let $S = S_{(m)} = k[x_1, \ldots, x_m]/(x_1, \ldots, x_m)^4$. Let $R = R_{(m)}$ be the trivial extension of S by its injective envelope (as described in [155, p. 70]). Then R is Gorenstein, generated by R_1, and $h(R) = \left(1, \binom{m+2}{3} + m, 2\binom{m+1}{2}, \binom{m+2}{3} + m, 1\right)$. Let x be an indeterminate of degree one. Choose $0 \neq t \in \operatorname{soc} R$, and define $R^{[1]} = R[x]/I$, where I is the ideal of $R[x]$ generated by xR_1 and $x^4 - t$. Then $R^{[1]}$ is Gorenstein, generated by $R_1^{[1]}$, and $h(R^{[1]}) = h(R) + (0, 1, 1, 1, 0)$. Thus defining $R^{[j]} = \left(R^{[j-1]}\right)^{[1]}$, we have for all $j \geq 1$ that $R^{[j]}$ is Gorenstein, generated by $R_1^{[j]}$, and $h\left(R^{[j]}\right) = h(R) + (0, j, j, j, 0)$. For any n, let m be the greatest integer such that $\binom{m+2}{3} + m \leq n$, and set $j = n - \binom{m+2}{3} - m$. Then the rings $T = R_{(m)}^{[j]}$ are Gorenstein, generated by T_1, and satisfy $h(T) = (1, n, t_n, n, 1)$, where

$$\lim_{n \to \infty} \frac{\log t_n}{\log n} = \frac{2}{3}.$$

(In fact, $\limsup t_n n^{-2/3} = \frac{3}{2} \cdot 6^{2/3}$.) This shows $\limsup \frac{\log f(n)}{\log n} \leq \frac{2}{3}$. But it follows easily just from the fact that $(1, n, f(n), n, 1)$ is an M-vector that $\liminf \frac{\log f(n)}{\log n} \geq \frac{2}{3}$. (In fact, $\liminf f(n)n^{-2/3} \geq \frac{1}{2} \cdot 6^{2/3}$.) Thus the proof follows. $\qquad\square$

It is open whether $f(n)n^{-2/3}$ converges to some limit c, which from the above argument must satisfy $\frac{1}{2} \cdot 6^{2/3} \leq c \leq \frac{3}{2} \cdot 6^{2/3}$. We conjecture $c = 6^{2/3}$. Facts (c) and (d) were first stated in [155, Ex. 4.3] (with the bound $c \leq 6^{2/3}$ due to an erroneous argument), but no proofs were provided. P. Kleinschmidt has subsequently shown [110] that

$$\limsup f(n)n^{-2/3} \leq 6^{2/3} ,$$

so if c exists then indeed $c \leq 6^{2/3}$.

Using the theory of ancestor ideals [101] [101, Lemma 1.1], one can give a purely linear algebraic definition (or characterization) of Gorenstein sequences. Let M_i denote the set of all monomials of degree i in the variables $x_1, \ldots, x_n$. Fix $s \in \mathbb{N}$ and a nonzero function $\sigma : M_s \to k$. For $0 \leq j \leq s$ define a matrix $A^{(j)}$, whose rows are indexed by M_j and columns by M_{s-j}, by the rule $A_{uv}^{(j)} = \sigma(uv)$. Let $h_j = \text{rank } A^{(j)}$. Then $(h_0, h_1, \ldots, h_s)$ is a Gorenstein sequence (over k) with $h_1 \leq n$, and all such Gorenstein sequences arise in this way.

7 Canonical modules of face rings

The next topic will be canonical modules of Cohen–Macaulay face rings. Recall (Theorem 12.8 of Chapter I) that $\Omega(R)$ is isomorphic to an ideal $I \subseteq R$ if and only if R is generically Gorenstein (i.e., $R_\wp$ is Gorenstein for all minimal primes $\wp$). Now, a face ring $k[\Delta]$ is generically a field. This suggests the following general problem: Imbed $\Omega(k[\Delta])$ as an ideal I of $k[\Delta]$, and describe $k[\Delta]/I$.

Recall that if R is a graded Cohen–Macaulay algebra of dimension d then $F(\Omega(R), \lambda) = (-1)^d F(R, 1/\lambda)$ (up to a shift in the grading of $\Omega(R)$). The right-hand side can be explicitly computed for any $R = k[\Delta]$.

7.1 Theorem. *Let Δ be any $(d-1)$-dimensional simplicial complex and give $k[\Delta]$ the fine grading. Then*

$$(-1)^d F\left(k[\Delta], 1/\lambda\right) = \sum_{F \in \Delta} (-1)^{d-|F|-1} \tilde{\chi}(\text{lk } F) \prod_{x_i \in F} \frac{\lambda_i}{1 - \lambda_i} .$$

Proof. $F(k[\Delta], \lambda) = \sum_{F \in \Delta} \prod_{x_i \in F} \frac{\lambda_i}{1-\lambda_i}$, so

$$(-1)^d F\left(k[\Delta], 1/\lambda\right) = (-1)^d \sum_{F \in \Delta} (-1)^{|F|} \prod_{x_i \in F} \frac{1}{1 - \lambda_i} .$$

Let $\alpha = (\alpha_1, \dots, \alpha_n) \in \mathbb{N}^n$ and $F = \{x_i \mid \alpha_i > 0\} \in \Delta$. The coefficient of λ^α is $(-1)^d \sum\limits_{\substack{G \in \Delta \\ G \supseteq F}} (-1)^{|G|} = (-1)^d \sum\limits_{G' \in \text{lk } F} (-1)^{|F|+|G'|} = (-1)^{d+|F|-1} \tilde{\chi}(\text{lk } F)$. $\square$

7.2 Corollary. *Let $|\Delta|$ be a manifold with boundary (possibly empty). Then*

$$(-1)^d F\left(k[\Delta], 1/\lambda\right) = (-1)^{d-1} \tilde{\chi}(\Delta) + \sum_{\substack{F \in \Delta - \partial \Delta \\ F \neq \emptyset}} \prod_{x_i \in F} \frac{\lambda_i}{1 - \lambda_i} .$$

If $|\Delta|$ is a Cohen–Macaulay manifold of dimension ≥ 1 with nonempty boundary then $\tilde{\chi}(\Delta) = 0$. Hence, in this case

$$F\left(\Omega(k[\Delta]), \lambda\right) = \sum_{F \in \Delta - \partial \Delta} \prod_{x_i \in F} \frac{\lambda_i}{1 - \lambda_i} = F(I, \lambda) ,$$

where I is the ideal of $k[\Delta]$ generated by all $F \in \Delta - \partial\Delta$. Thus it is natural to ask whether $\Omega(k[\Delta]) \cong I$.

7.3 Theorem (Hochster, unpublished). *Let $|\Delta|$ be a Cohen–Macaulay manifold with nonempty boundary $|\partial\Delta|$. Then*

$$I \cong \Omega(k[\Delta]) \Leftrightarrow \partial\Delta \text{ is Gorenstein} .$$

(**Remark:** $\partial\Delta$ is Gorenstein e.g. if $|\Delta|$ is orientable.)

Proof. ($\Leftarrow$) Let C be a cone over $\partial\Delta$ with vertex y. Let $\Gamma = \Delta \cup C$, where Δ and C are identified along $\partial\Delta = \partial C$. $|\Gamma|$ is a manifold, except possibly at y where $\text{lk } y = \partial\Delta$. By an elementary application of the Mayer-Vietoris exact sequence, $\widetilde{H}_d(\Gamma) \cong \widetilde{H}_{d-1}(\partial\Delta) \cong k$ and $H_i(\Gamma) = 0$ for $i \neq d$. One similarly checks the proper links to find that Γ is Gorenstein. Since $S = k[\Gamma]$ is Gorenstein and $R = k[\Delta] = k[\Gamma]/J$ is Cohen–Macaulay of the same dimension d it follows that $\Omega(k[\Delta]) \cong \text{Hom}_S(R, S)$ (it is a well-known fact that in general if S is Gorenstein and $R = S/J$ then $\Omega(R) \cong \text{Ext}_{\dim S - \dim R}(R, S)$). Thus, $\Omega(k[\Delta]) \cong \text{Ann}_S J = I$.

($\Rightarrow$) Recall (Theorem 12.9 of Chapter I) that in general if R is Cohen–Macaulay and if an ideal I is isomorphic to $\Omega(R)$, then R/I is Gorenstein. In the present situation $k[\Delta]/I \cong k[\partial\Delta]$, so if $I \cong \Omega(k[\Delta])$ then $\partial\Delta$ is Gorenstein. $\square$

What can one say about canonical modules of manifolds without boundary? If $|\Delta|$ is a Cohen–Macaulay *orientable* manifold without boundary, then $k[\Delta]$ is Gorenstein so $\Omega(k[\Delta]) \cong k[\Delta]$. If $|\Delta|$ is a Cohen–Macaulay *nonorientable* manifold without boundary, then $\tilde{\chi}(\Delta) = 0$ so by our earlier computation $F(\Omega(k[\Delta]), \lambda) = F(k[\Delta]_+, \lambda)$ where $k[\Delta]_+ = (x_1, x_2, \ldots, x_n)$. One might first suspect that $\Omega(k[\Delta]) \cong k[\Delta]_+$, but this would contradict Theorem 12.9 of Chapter I. The actual description of $\Omega(k[\Delta])$ can be obtained as follows. Orient $\operatorname{st} x_i = \{F \in \Delta \mid F \cup x_i \in \Delta\}$ for each $x_i \in V$. Define the 1-Cech cocycle σ as follows. If $\{x_i, x_j\} \in \Delta$, let

$$
\sigma\{x_i, x_j\} = \begin{cases} 1, & \text{if } \operatorname{st} x_i \text{ and } \operatorname{st} x_j \text{ have compatible orientations} \\ & \quad \text{(i.e., agree on intersection)}, \\ -1, & \text{if not}. \end{cases}
$$

(**Remark:** σ is the obstruction cocycle for orientability; the image of σ in $H^1(\Delta)$ is 0 if and only if Δ is orientable.) Paste together the ideals $x_i k[\Delta]$ as follows: identify $x_i x_j k[x] \subseteq x_i k[\Delta]$ with $\sigma\{x_i, x_j\} x_i x_j k[\Delta] \subseteq x_j k[\Delta]$. In other words, form $M = \coprod_i x_i k[\Delta]/(x_j e_i - \sigma\{x_i, x_j\} x_i e_j)$ where e_i is the image of x_i in the ith summand.

7.4 Theorem (Hochster, unpublished). $M \cong \Omega(k[\Delta])$.

The proof is omitted.

The problem of giving a "nice" description of $\Omega(k[\Delta])$ for arbitrary Cohen–Macaulay Δ was solved by Gräbe [84], who also refined Theorem 4.1 by giving the structure of $H^i(k[\Delta])$ as a $k[x_1, \ldots, x_n]$-module. We will only state a special case of Gräbe's result due primarily to Baclawski. Define Δ to be *doubly Cohen–Macaulay* (2-Cohen–Macaulay) if Δ is Cohen–Macaulay, and for every vertex x of Δ the subcomplex $\Delta\backslash x = \{F \in \Delta : x \notin F\}$ is Cohen–Macaulay of the same dimension as Δ. Walker [181] has shown that double-Cohen–Macaulayness is a topological property, i.e., depends only on $|\Delta|$. For instance, spheres are 2-Cohen–Macaulay but cells are not. If Δ is Cohen–Macaulay, then Δ is 2-Cohen–Macaulay if and only if $(-1)^{d-1}\tilde{\chi}(\Delta) = \text{type } k[\Delta]$ (see [9]). For any Δ we may identify $F \in \Delta$ with $\prod_{x \in F} x \in k[\Delta]$. In this way the augmented oriented chain complex of Δ (over k) can be imbedded (as a vector space) in $k[\Delta]$. In particular, if $\dim \Delta = d - 1$ then the reduced homology group $\widetilde{H}_{d-1}(\Delta; k)$ is imbedded in $k[\Delta]$, since it is a subspace of the $(d-1)$-chains. The following result was first proved by Baclawski [10] for balanced complexes (defined in Chapter 3.4). A proof of the general case appears in [48, Cor. 5.6.7].

7.5 Theorem. *Suppose Δ is 2-Cohen–Macaulay of dimension $d - 1$. Then $\Omega(k[\Delta])$ is isomorphic to the ideal of $k[\Delta]$ generated by $\widetilde{H}_{d-1}(\Delta; k)$.*

$\square$

If M is any Cohen–Macaulay module then $\Omega^2(M) := \Omega(\Omega(M))$ is isomorphic to M. The non-Cohen–Macaulay case was worked out by Hochster. Here we define $\Omega(M) = \operatorname{Ext}_A^{n-d}(M, A)$, where M is finitely-generated over the Gorenstein ring A, and where $n = \dim A, d = \dim M$. For face rings $k[\Delta]$ it has the following informal description. First "purify" Δ by removing all faces not contained in a face of maximum dimension $d - 1$. This yields a pure simplicial complex Δ'. Next choose a nonempty face F of Δ' of smallest possible dimension $\leq d - 3$ such that lk F is not connected, and "pull apart" Δ' at F by creating a copy of F_i of F for each connected component K_i of lk F, so that lk $F_i = K_i$.

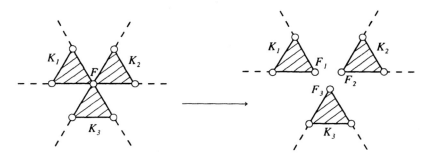

Find another nonempty face F' of smallest possible dimension $\leq d - 3$ such that lk F' is not connected, and repeat the procedure. Continue until the link of every nonempty face of dimension $\leq d - 3$ is connected. Let $\Delta_1, \Delta_2, \ldots, \Delta_j$ be the connected components of the resulting simplicial complex. Then

$$\Omega^2(k[\Delta]) \cong \coprod_{i=1}^{j} k[\Delta_i],$$

as a $k[x_1, \ldots, x_n]$-module or $k[\Delta]$-module. In particular, $\Omega^2(k[\Delta]) \cong k[\Delta]$ if and only if Δ is pure and the link of every face (including $\emptyset$) of dimension $\leq d - 3$ is connected.

8 Buchsbaum complexes

The final topic of this chapter will be a brief glimpse of Buchsbaum complexes. Let R be an $\mathbb{N}$-graded finitely-generated algebra, and let M be a d-dimensional $\mathbb{Z}$-graded finitely-generated R-module. M is said to be *Buchsbaum* if for every h.s.o.p. $\theta_1, \ldots, \theta_d$ and all $1 \leq i \leq d$,

$$\{u \in M/(\theta_1 M + \cdots + \theta_{i-1} M) \mid u\theta_i = 0\} = \operatorname{soc} M/(\theta_1 M + \cdots + \theta_{i-1} M) \ .$$

When applied to face rings $k[\Delta]$ this notion carries over to simplicial complexes Δ. A comprehensive general reference is [176].

8.1 Theorem (Schenzel [142]). *Let Δ be a finite simplicial complex and k a field. Then the following are equivalent:*

(i) Δ *is Buchsbaum over* k,

(ii) Δ *is pure, and* $k[\Delta]_\wp$ *is Cohen–Macaulay for all primes* $\wp \neq k[\Delta]_+$,

(iii) Δ *is pure, and for all* $F \in \Delta$, *the conditions* $F \neq \emptyset$ *and* $i < \dim(\operatorname{lk} F)$ *imply* $\widetilde{H}_i(\operatorname{lk} F; k) = 0$,

(iv) *for all* $p \in X = |\Delta|$ *and* $i < \dim X$, $H_i(X, X - p; k) = 0$,

(v) $\dim_k H^i(k[\Delta]) < \infty$ *if* $0 \leq i < \dim k[\Delta] = d$ *(in which case* $H^i(k[\Delta]) \cong \widetilde{H}_{i-1}(\Delta; k)$, $i < d$).

Remark: Schenzel omitted the condition that Δ is pure in (ii) and (iii), as was pointed out by Miyazaki [128, Remark, p. 251].

Remark: A characterization such as (v) is not known for general Buchsbaum rings R. In general, a necessary but not sufficient condition for M to be Buchsbaum is that $\dim_k H^i(M) < \infty$ for $0 \leq i < \dim M$.

8.2 Theorem (Schenzel [142]). *Let Δ be a Buchsbaum complex, let* $\deg x_i = 1$, *and let* $\theta_1, \ldots, \theta_d \in k[\Delta]_1$ *be an h.s.o.p. Then*

$$(1 - \lambda)^d F(k[\Delta], \lambda) = F(k[\Delta]/(\theta_1, \ldots, \theta_d), \lambda)$$
$$- \sum_{j=1}^{d} \binom{d}{j} \left(\sum_{i=0}^{j-1} (-1)^{j-i-1} \dim_k \widetilde{H}_{i-1}(\Delta) \right) \lambda^j \ .$$

The last term can be interpreted as measuring the error when we leave the Cohen–Macaulay case. Set

$$h_0 + h_1\lambda + \cdots + h_d\lambda^d := (1 - \lambda)^d F(k[\Delta], \lambda) , \quad \text{and}$$
$$g_0 + g_1\lambda + \cdots + g_d\lambda^d := F(k[\Delta]/(\theta_1, \ldots, \theta_d), \lambda) .$$

Then $g_1 = n - d$ and $g_j \leq \binom{n-d+j-1}{j}$, the number of monomials of degree j in the $n - d$ variables. Hence,

$$h_j \leq \binom{n-d+j-1}{j} - (-1)^j \binom{d}{j} \sum_{i=-1}^{j-2} (-1)^i \dim_k \widetilde{H}_i(\Delta) . \qquad (5)$$

This inequality for the h-vector of a Buchsbaum complex Δ generalizes the inequality for a Cohen–Macaulay complex which yields the Upper Bound Conjecture. (It should be noted, however, that the Upper Bound Conjecture does not follow formally from (5); one also needs the Dehn–Sommerville equations.) It implies the following bounds on the f-vector $f(\Delta) = (f_0, \ldots, f_{d-1})$:

$$f_j \leq \binom{n}{j} - \binom{d}{j} \sum_{i=-1}^{j-2} \binom{j-1}{i+1} \dim_k \widetilde{H}_i(\Delta) .$$

If in addition to being Buchsbaum Δ is an orientable homology manifold (i.e., $\widetilde{H}_{\dim(\mathrm{lk}\,F)}(\mathrm{lk}\,F) \cong k$ for all $F \in \Delta$), then

$$h_i = h_{d-i} + (-i)^d \binom{d}{i} \left((-1)^{d-1} - \tilde{\chi}(\Delta) \right) .$$

Note that if d is even then $\tilde{\chi}(\Delta) = -1$ due to Poincaré duality, hence in this case $h_i = h_{d-i}$. Recently I. Novik [133] has used equation (5) and some further properties of Buchsbaum rings to obtain some significant generalizations of the Upper Bound Conjecture for spheres.

Chapter III

Further Aspects of Face Rings

In this chapter we will briefly survey some additional topics related to combinatorics and commutative algebra, mostly dealing with the face ring of a simplicial complex. Our main focus will be on properties of face rings which have applications to combinatorics.

1 Simplicial polytopes, toric varieties, and the g-theorem

There is a special class of triangulations of spheres for which the Upper Bound Conjecture (or more accurately, the Upper Bound Theorem) for spheres (Corollary II.4.5) can be greatly strengthened to a complete characterization of the h-vector. We will be rather brief in this presentation since the proofs involve machinery outside of commutative algebra. We assume basic knowledge of convex polytopes such as discussed in Chapter 0. For surveys of the topic of this section, see [16], [77, §5.6], [161].

First we state the theorem whose proof we will sketch. A convex polytope $\mathcal{P}$ is *simplicial* if every proper face is a simplex. For instance, the tetrahedron, octahedron, and icosahedron are simplicial, but not the cube or dodecahedron. The boundary $\partial\mathcal{P}$ of a simplicial d-polytope $\mathcal{P}$ is the geometric realization of a $(d-1)$-dimensional simplicial complex $\Delta = \Delta_\mathcal{P}$, called the *boundary complex* of $\mathcal{P}$. In other words, the vertices of Δ are just the vertices of $\mathcal{P}$, and a set F of vertices of Δ forms a face if and only if they are the vertices of a face of $\mathcal{P}$ other than $\mathcal{P}$ itself. Since $\partial\mathcal{P}$ is a geometric realization of Δ it follows that $|\Delta| \approx \mathbb{S}^{d-1}$, so Corollaries II.4.4 and II.4.5 apply to Δ. Moreover, since spheres are Gorenstein by Theorem II.5.1, it follows that Δ satisfies the Dehn-Sommerville equations (Chapter II.6), i.e., the h-vector of Δ satisfies $h_i = h_{d-i}$. We will also call the h-vector $h(\Delta)$ the h-vector of $\mathcal{P}$.

The main result of this section is Theorem 1.1 below. It was conjectured in 1971 by Peter McMullen [123]. The conjecture was known as the *g-conjecture* because of McMullen's use of $g_i^{(d+1)}$ for our $h_i - h_{i-1}$. The implication (a) $\Rightarrow$ (b) was proved by Stanley [159], while (b) $\Rightarrow$ (a) was shown by Billera and Lee [20], [21]. Thus the result is now known as the *g-theorem* (for simplicial polytopes). For more historical information, see [161, p. 221] and [170, p. 318].

1.1 Theorem. *Let* $\mathbf{h} = (h_0, h_1, \ldots, h_d) \in \mathbb{Z}^{d+1}$. *The following two conditions are equivalent.*

(a) *There exists a simplicial d-polytope* $\mathcal{P}$ *such that* $h(\mathcal{P}) = \mathbf{h}$.

(b) $h_0 = 1$, $h_i = h_{d-i}$ *for all* i, *and* $(h_0, h_1 - h_0, h_2 - h_1, \ldots, h_{\lfloor d/2 \rfloor} - h_{\lfloor d/2 \rfloor - 1})$ *is an M-vector (as defined in Chapter II.2).*

We will not discuss the implication (b) $\Rightarrow$ (a), which is proved by an ingenious construction. The main tool used to prove (a) $\Rightarrow$ (b) is the theory of toric varieties. Let $\mathcal{P}$ be a d-dimensional convex polytope in $\mathbb{R}^d$ with integer vertices. (More generally, the vertices can be in a lattice L, but we lose nothing here by taking $L = \mathbb{Z}^d$.) We call $\mathcal{P}$ an *integral convex polytope*. If $\alpha = (\alpha_1, \ldots, \alpha_d) \in \mathbb{Z}^d$ then write $x^\alpha = x_1^{\alpha_1} \cdots x_d^{\alpha_d}$. Let $\mathbb{C}^* = \mathbb{C} - \{0\}$. Define

$$X_{\mathcal{P}} = \mathrm{cl}\left\{(x^{\alpha^1}y, \ldots, x^{\alpha^n}y) : x_1, \ldots x_d, y \in \mathbb{C}^*\right\} \subset P_{\mathbb{C}}^{n-1}, \qquad (1)$$

where the vertex set of $\mathcal{P}$ is $\{\alpha^1, \ldots, \alpha^n\}$, where $P_{\mathbb{C}}^{n-1}$ denotes $(n-1)$-dimensional complex projective space, and where cl denotes closure (either in the Zariski topology or classical topology — they coincide here). The set $X_{\mathcal{P}}$ is a complex projective variety, called a (projective) *toric variety*. We are only considering a special case of the general notion of toric variety and have given a "naive" definition which is not so readily adaptable to the general case. For further information on toric varieties see [60], [73], [77], [134].

1.2 Example. Let $\mathcal{P}$ be the square with vertices $(0,0)$, $(1,0)$, $(0,1)$, and $(1,1)$. Then

$$X_{\mathcal{P}} = \mathrm{cl}\left\{(y, x_1 y, x_2 y, x_1 x_2 y) : x_1, x_2, y \in \mathbb{C}^*\right\} \subset P_{\mathbb{C}}^3.$$

For any $a, b \in \mathbb{C}^*$ we can let $x_1 = a/y$ and $x_2 = b/a$. Letting $y \to 0$ shows that $(0, a, 0, b) \in X_{\mathcal{P}}$. We can then let $a \to 0$ or $b \to 0$ to get

$(0, a, 0, b) \in X_{\mathcal{P}}$ unless $a = b = 0$. This illustrates the meaning of the closure operator cl. Denoting the coordinates of $P_{\mathbb{C}}^3$ by $z_0, \ldots, z_3$, it is easy to see that $X_{\mathcal{P}}$ is defined by the single equation $z_0 z_3 - z_1 z_2 = 0$. The variety $X_{\mathcal{P}}$ is isomorphic to the two-dimensional complex torus $P_{\mathbb{C}}^1 \times P_{\mathbb{C}}^1$, the isomorphism given by $((s_0, s_1), (t_0, t_1)) \mapsto (s_0 t_0, s_0 t_1, s_1 t_0, s_1 t_1)$. (This is nothing but the usual Segre embedding $P_{\mathbb{C}}^1 \times P_{\mathbb{C}}^1 \to P_{\mathbb{C}}^3$.) Let $\beta_i(X) = \dim_{\mathbb{R}} H^i(X; \mathbb{R})$, the ith betti number of the topological space X (over the reals). Since $\sum \beta_i(P_{\mathbb{C}}^1) x^i = 1 + x^2$, there follows $\sum \beta_i(X_{\mathcal{P}}) x^i = (1 + x^2)^2 = 1 + 2x^2 + x^4$. Note that the boundary of $\mathcal{P}$ is the geometric realization of a simplicial complex with h-vector $(1, 2, 1)$. This connection between the cohomology of $X_{\mathcal{P}}$ and the h-vector of the boundary complex of $\mathcal{P}$ is of course not a coincidence and is a basic fact which we will need in applying toric varieties to the combinatorics of $\mathcal{P}$.

Let $\mathcal{P}$ be an integral convex d-polytope in $\mathbb{R}^d$, and let $X_{\mathcal{P}}$ be the corresponding toric variety. It is easy to see that $X_{\mathcal{P}}$ is a complex projective variety of (complex) dimension d (so real dimension $2d$). Let $H^*(X_{\mathcal{P}}; \mathbb{R})$ denote the singular cohomology ring of $X_{\mathcal{P}}$ over $\mathbb{R}$. Thus we have a natural grading

$$H^*(X_{\mathcal{P}}; \mathbb{R}) = H^0(X_{\mathcal{P}}) \oplus H^1(X_{\mathcal{P}}) \oplus \cdots \oplus H^{2d}(X_{\mathcal{P}}),$$

where each $H^i(X_{\mathcal{P}})$ is a finite-dimensional vector space over $\mathbb{R}$. A fundamental result on toric varieties is the following [60, Thm. 10.8], [77, §5.2]. Here and in what follows we use l.s.o.p. as an abbreviation for "linear system of parameters," i.e., a homogeneous system of parameters all of degree one.

1.3 Theorem. *Let $\mathcal{P}$ be an integral simplicial d-polytope in $\mathbb{R}^d$ with boundary complex Δ, and let $X_{\mathcal{P}}$ be the corresponding toric variety. Then there is an algebra isomorphism*

$$\varphi : H^*(X_{\mathcal{P}}; \mathbb{R}) \to \mathbb{R}[\Delta]/(\theta_1, \ldots, \theta_d),$$

for a certain l.s.o.p. $\theta_1, \ldots, \theta_d$ of $\mathbb{R}[\Delta]$. This isomorphism halves degree, so $\varphi(H^{2i}(X_{\mathcal{P}})) = \mathbb{R}[\Delta]_i$, and $H^{2i+1}(X_{\mathcal{P}}) = 0$.

We need one further result concerning the cohomology ring $H^*(X_{\mathcal{P}}; \mathbb{R})$ when $\mathcal{P}$ is an integral simplicial polytope. This result is known as the *hard Lefschetz theorem*. Classically the hard Lefschetz theorem was proved for smooth varieties. Although the varieties $X_{\mathcal{P}}$ are not in general smooth, their singularities (when $\mathcal{P}$ is simplicial) are so nice (namely, they are "finite quotient singularities") that their cohomology "acts like" that of a smooth variety. For instance, $X_{\mathcal{P}}$ satisfies Poincaré duality, which in light of Theorem 1.3 amounts to saying that the ring $\mathbb{R}[\Delta]/(\theta_1, \ldots, \theta_d)$ is Gorenstein.

1.4 Theorem. *Preserve the notation of Theorem 1.3. Then there exists an element $\omega \in H^2(X_{\mathcal{P}})$ (namely, the class of a hyperplane section with respect to a projective embedding of $X_{\mathcal{P}}$; for the canonical embedding (1) used to define $X_{\mathcal{P}}$ this element is just $x_1 + x_2 + \cdots + x_n$, the sum of all vertices of Δ) such that for all $0 \le i \le \lfloor d/2 \rfloor$, the map $\omega^{d-2i} : H^{2i}(X_{\mathcal{P}}) \to H^{2d-2i}(X_{\mathcal{P}})$ (given by multiplication by ω^{d-i}) is a bijection.*

It is now easy to prove the implication (a) $\Rightarrow$ (b) of Theorem 1.1. The condition $h_0 = 1$ is trivial, while $h_i = h_{d-i}$ has already been discussed (and is not difficult to prove by elementary reasoning). It remains to show that $(h_0, h_1 - h_0, h_2 - h_1, \ldots, h_{\lfloor d/2 \rfloor} - h_{\lfloor d/2 \rfloor - 1})$ is an M-vector. Preserve the above notation, and write $S = \mathbb{R}[\Delta]/(\theta_1, \ldots, \theta_d) \cong H^*(X_{\mathcal{P}}; \mathbb{R})$. Since S is Cohen–Macaulay (by Corollary II.4.4) we have $\dim_{\mathbb{R}} S_i = h_i(\Delta)$. By Theorem 1.3 we can identify S_i with $H^{2i}(X_{\mathcal{P}})$. It follows easily from Theorem 1.4 that the map $\omega : S_{i-1} \to S_i$ is injective for $0 \le i \le \lfloor d/2 \rfloor$. Hence the Hilbert function of the quotient ring $S/\omega S$ satisfies for $0 \le i \le \lfloor d/2 \rfloor$ the condition that

$$
\begin{aligned}
H(S/\omega S, i) &= H(S, i) - H(\omega S, i) \\
&= H(S, i) - H(S, i-1) \\
&= h_i - h_{i-1}.
\end{aligned}
$$

We have found a graded algebra $S/\omega S$, generated in degree one, whose Hilbert function satisfies $H(S/\omega S, i) = h_i - h_{i-1}$ for $0 \le i \le \lfloor d/2 \rfloor$. Hence by Corollary II.2.4 the proof is complete.

Ever since the above proof appeared it became a challenge to find a more elementary proof, in particular, a proof avoiding toric varieties. Such a proof was finally found by McMullen [125] (corrected and simplified in [126]), though it is still a difficult proof.

2   Shellable simplicial complexes

A shellable simplicial complex is a special kind of Cohen–Macaulay complex with a simple combinatorial definition. Shellability is a simple but powerful tool for proving the Cohen–Macaulay property, and almost all

Cohen–Macaulay complexes arising "in nature" turn out to be shellable. Moreover, a number of invariants associated with Cohen–Macaulay complexes can be described more explicitly or computed more easily in the shellable case. In particular, the h-vector and the basis elements $\eta_1, \ldots, \eta_t$ of Theorem I.5.10 have direct combinatorial descriptions.

2.1 Definition. A simplicial complex Δ is *pure* if each facet (= maximal face) has the same dimension. We say that a pure simplicial complex Δ is *shellable* if its facets can be ordered $F_1, F_2, \ldots, F_s$ such that the following condition holds. Write Δ_j for the subcomplex of Δ generated by $F_1, \ldots, F_j$, i.e.,

$$\Delta_j = 2^{F_1} \cup 2^{F_2} \cup \cdots \cup 2^{F_j},$$

where $2^F = \{G : G \subseteq F\}$. Then we require that for all $1 \leq i \leq s$, the set of faces of Δ_i which do not belong to Δ_{i-1} has a unique minimal element (with respect to inclusion). (When $i = 1$, we have $\Delta_0 = \emptyset$ and $\Delta_1 = 2^{F_1}$, so $\Delta_1 - \Delta_0$ has the unique minimal element $\emptyset$.) The linear order $F_1, \ldots, F_s$ is then called a *shelling order* or a *shelling* of Δ.

There are several equivalent ways to define shellability. For instance, if $\dim(\Delta) = d - 1$, then $F_1, \ldots, F_s$ is a shelling if for all $2 \leq i \leq s$, the facet F_i is attached to Δ_{i-1} on a union of $(d-2)$-dimensional faces of F_i. (In other words, the subcomplex $\Delta_{i-1} \cap 2^{F_i}$ is pure of dimension $d - 2$.)

2.2 Example. Let Δ be the simplicial complex with facets $F = abc$ (short for $\{a, b, c\}$), $G = bcd$, and $H = cde$. The order F, G, H is a shelling, since the unique minimal face of $\Delta_1 - \Delta_0$ is $\emptyset$, of $\Delta_2 - \Delta_1$ is d, and of $\Delta_3 - \Delta_2$ is e. On the other hand, F, H, G is not a shelling, since $\Delta_2 - \Delta_1 = \{d, e, cd, ce, de, cde\}$, with minimal elements d and e. Equivalently, H is attached to $\Delta_1 = 2^F$ on the face c, which has dimension 0, not 1. See Figure 3.1.

It is obvious that every shellable simplicial complex of dimension at least one is connected. An example of a connected pure nonshellable simplicial complex is given by the facets abc and cde.

Given a shelling $F_1, \ldots, F_s$ of Δ, define the *restriction* $r(F_i)$ of F_i to be the unique minimal element of $\Delta_i - \Delta_{i-1}$. If $G \subseteq F$, write

$$[G, F] = \{H : G \subseteq H \subseteq F\},$$

the (closed) *interval* from G to F. It is an immediate consequence of the definition of a shelling that

$$\Delta = [r(F_1), F_1] \cup \cdots \cup [r(F_s), F_s] \quad \text{(disjoint union)}. \tag{2}$$

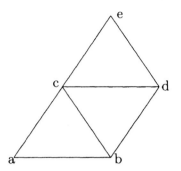

Figure 3.1: A shellable simplicial complex

This leads us to define a pure simplicial complex Δ to be *partitionable* if Δ can be written as a disjoint union

$$\Delta = [G_1, F_1] \cup \cdots \cup [G_s, F_s], \tag{3}$$

where each $F_\imath$ is a facet of Δ. We then call the right-hand side of (3) a *partitioning* of Δ (called a "facet cover" in [111]). Equation (2) shows that shellable simplicial complexes are partitionable.

2.3 Proposition. *Let (3) be a partitioning of Δ. Let $(h_0, h_1, \ldots, h_d)$ denote the h-vector of Δ. Then*

$$h_\imath = \#\{j : |G_\jmath| = i\}.$$

Equivalently,

$$\sum_{\imath=0}^{d} h_\imath x^\imath = \sum_{\jmath=1}^{s} x^{|G_\jmath|}.$$

Proof. Let $(f_0, f_1, \ldots, f_{d-1})$ denote the f-vector of Δ (with $f_{-1} = 1$ unless $\Delta = \emptyset$). The definition of the h-vector given in Chapter II.2 is equivalent to the identity

$$\sum_{\imath=0}^{d} f_{i-1}(x-1)^{d-\imath} = \sum_{\imath=0}^{d} h_\imath x^{d-i}. \tag{4}$$

(In general, equation (4) seems to be the most useful and convenient way to express the relationship between the f-vector and h-vector.) The left-hand side of (4) may be rewritten as $\sum_{F\in\Delta}(x-1)^{d-|F|}$. Hence given the partitioning (3), we have

$$\sum_{i=0}^{d} h_i x^{d-i} = \sum_{j=1}^{s} \sum_{G_j \subseteq F \subseteq F_j} (x-1)^{d-|F|}.$$

If $|G_j| = i$ then

$$\sum_{G_j \subseteq F \subseteq F_j} (x-1)^{d-|F|} = \sum_{k=0}^{d-i} \binom{d-i}{k} (x-1)^{d-i-k} \tag{5}$$

$$= x^{d-i}, \tag{6}$$

and the proof follows. $\square$

We next want to show that there is a choice of the basis elements η_i of Theorem I.5.10 with a simple explicit description when M is the face ring of a shellable simplicial complex. First we need a condition on when a sequence $\theta_1, \ldots, \theta_d$ is an l.s.o.p. for $k[\Delta]$. While in this section we only need the easy direction of Lemma 2.4(a) below (the "only if" part), the full result will be of great importance in subsequent sections. A general theme of this chapter will be to obtain extra information from $k[\Delta]$ by choosing *special* l.s.o.p.'s, and for this we need to recognize an l.s.o.p.

Given an element $\theta = \sum_i \alpha_i x_i \in k[\Delta]_1$ and a face $F \in \Delta$, define the *restriction* of θ to F, denoted $\theta|_F$, by

$$\theta|_F = \sum_{x_i \in F} \alpha_i x_i.$$

If F is a face of Δ, define the *face monomial*

$$x^F = \prod_{x_i \in F} x_i. \tag{7}$$

The next result was first given (in a slightly weaker form) by Kind and Kleinschmidt [108]. Part (a) was also stated without proof in [156, Rmk. on p. 150]. See also [81, Thm. 4.3].

2.4 Lemma. (a) *Let $k[\Delta]$ be a face ring of Krull dimension d, and let $\theta_1, \ldots, \theta_d \in k[\Delta]_1$. Then $\theta_1, \ldots, \theta_d$ is an l.s.o.p. for $k[\Delta]$ if and only if for every face F of Δ (or equivalently, for every facet F of Δ), the restrictions $\theta_1|_F, \ldots, \theta_d|_F$ span a vector space of dimension equal to $|F|$.*

(b) *If $\theta_1, \ldots, \theta_d$ is an l.s.o.p. for $k[\Delta]$, then the quotient ring*

$$k[\Delta]/(\theta_1, \ldots, \theta_d)$$

is spanned (as a vector space over k) by the face monomials $x^F, F \in \Delta$.

Proof. Let $S = k[\Delta]/(\theta_1, \ldots, \theta_d)$. If $\theta_1, \ldots, \theta_d$ is an l.s.o.p. for $k[\Delta]$, then S is a finite-dimensional vector space. Fix a face $F \in \Delta$. S remains finite-dimensional when we kill all vertices of Δ not in F. Thus $\theta_1|_F, \ldots, \theta_d|_F$ must span all homogeneous elements of degree one in the polynomial ring $k[F]$, proving the "only if" part of (a).

Now assume that $\theta_1, \ldots, \theta_d$ satisfies the condition in (a) on the restrictions $\theta_1|_F, \ldots, \theta_d|_F$. To complete the proof of both (a) and (b) it suffices to show that S is spanned by face monomials, since there are only finitely many such monomials. We first claim that if F is a facet of Δ, then $x^F \in \mathrm{soc}(S)$, where soc is defined preceding Theorem I.12.4. We need to show that for every vertex x_i of Δ, we have $x_i x^F = 0$ in S. If $x_i \notin F$ then in fact $x_i x^F = 0$ in $k[\Delta]$, so assume $x_i \in F$. By the assumption on $\theta_1, \ldots, \theta_d$, some linear combination ψ of them has the form

$$\psi = x_i + \sum_{x_j \notin F} \alpha_j x_j.$$

Then in S we have $x_i x^F = (x_i - \psi)x^F = 0$, so $x^F \in \mathrm{soc}(S)$ as claimed.

Now let I be the ideal of S generated by all x^F where F is a facet of Δ, and let $S' = S/I$. Since the generators x^F of I all belong to $\mathrm{soc}(S)$, I is in fact spanned as a vector space by the face monomials x^F, where F is a facet of Δ. Hence $\dim S = \dim S'$, where dim denotes Krull dimension. Let Δ' be Δ with all its facets removed, and let I' be the ideal of S' generated by all x^F where F is a facet of Δ'. Exactly the same reasoning as before shows that $\dim S = \dim S' = \dim S'/I'$. Continuing in this way, we get $\dim S = \dim S/J$, where J is the ideal generated by *all* face monomials x^F, $F \in \Delta$. Since $S/J = 0$, we have $\dim S = 0$. Hence $\theta_1, \ldots, \theta_d$ is an l.s.o.p. for $k[\Delta]$. □

2.5 Theorem. *Let $F_1, \ldots, F_s$ be a shelling of the simplicial complex Δ, and let $\theta_1, \ldots, \theta_d$ be an l.s.o.p. for $k[\Delta]$. Let*

$$B = \left\{ x^{r(F_i)} : 1 \leq i \leq s \right\}.$$

Then $k[\Delta]$ is Cohen–Macaulay, and B is a k-basis for $S = k[\Delta]/(\theta_1, \ldots, \theta_d)$. Equivalently (by Theorem I.5.10), $k[\Delta]$ is a free module over the polynomial ring $k[\theta_1, \ldots, \theta_d]$ with basis B.

Proof. One can show that $k[\Delta]$ is Cohen–Macaulay by a simple Mayer–Vietoris argument together with Reisner's theorem (Theorem II.4.2). There is also a simple inductive proof [149] *via* commutative algebra (inspired by [94] and repeated in [48, Thm. 5.1.13]). We will give a more combinatorial argument, based upon the following elementary result (see [155, Cor. 3.2]).

2.6 Lemma. *Let* $\theta_1, \ldots, \theta_d$ *be homogeneous elements of positive degree of a finitely-generated $\mathbb{N}$-graded k-algebra R (or even more generally, a finitely-generated $\mathbb{Z}$-graded R-module). Let $f_i = \deg \theta_i$, and let $S = R/(\theta_1, \ldots, \theta_d)$. If $F(\lambda) = \sum a_i \lambda^i$ and $G(\lambda) = \sum b_i \lambda^i$ are two power series with real coefficients, then write $F(\lambda) \le G(\lambda)$ to mean $a_i \le b_i$ for all i. Then*

$$F(R, \lambda) \le \frac{F(S, \lambda)}{\prod_{i=1}^{d}(1 - \lambda^{f_i})},$$

with equality if and only if $\theta_1, \ldots, \theta_d$ is an R-sequence.

Returning to the proof of Theorem 2.5, we have by Lemma 2.6 and the definition of the h-vector $h(\Delta) = (h_0, \ldots, h_d)$ following the proof of Corollary II.2.5 that

$$\frac{h_0 + h_1 \lambda + \cdots + h_d \lambda^d}{(1 - \lambda)^d} \le \frac{F(S, \lambda)}{(1 - \lambda)^d}.$$

Since by Proposition 2.3 we have $\sum_{x^F \in B} \lambda^{\#F} = h_0 + h_1 \lambda + \cdots + h_d \lambda^d$, it follows that if B spans S then $\theta_1, \ldots, \theta_d$ is an R-sequence and that B is a k-basis for S. Hence it suffices to prove that B spans S.

The proof is now completely analogous to the proof of Lemma 2.4(b). We use induction on s, the assertion being clear for $s = 0$ (or even $s = 1$). We first claim that $x^{r(F_s)} \in \mathrm{soc}(S)$. If $x_i \notin F_s$ then $\{x_i\} \cup r(F_s) \notin \Delta$ by the definition of shelling. Hence $x_i x^{r(F_s)} = 0$ in $k[\Delta]$, so also in S. If $x_i \in F_s$ then there is some linear combination ψ of $\theta_1, \ldots, \theta_d$ of the form

$$\psi = x_i + \sum_{x_j \notin F_s} \alpha_j x_j.$$

Then just as in the proof of Lemma 2.4 we have $x_i x^{r(F_s)} = (x_i - \psi) x^{r(F_s)} = 0$, so $x^{r(F_s)} \in \mathrm{soc}(S)$ as claimed.

Now let $S' = S/x^{r(F_s)} S$; and let $\Delta_{s-1} = 2^{F_1} \cup \cdots \cup 2^{F_{s-1}}$, the simplicial complex generated by $F_1, \ldots, F_{s-1}$. We may regard $k[\Delta_{s-1}]$ as a $k[\Delta]$-module via the isomorphism $k[\Delta_{s-1}] \cong k[\Delta]/(x^{r(F_s)})$. Then $\theta_1, \ldots, \theta_d$ is also an l.s.o.p. for $k[\Delta_{s-1}]$, and $S' \cong k[\Delta_{s-1}]/(\theta_1, \ldots, \theta_d)$. By the induction hypothesis we have that $x^{r(F_1)}, \ldots, x^{r(F_{s-1})}$ spans S', so $x^{r(F_1)}, \ldots, x^{r(F_s)}$ spans S. The proof follows by induction. $\square$

In the next two sections we will see some interesting examples of shellable simplicial complexes. Here are some miscellaneous results and problems concerning shellability.

- Define a topological space X to be *Cohen–Macaulay* if it is the geometric realization of a Cohen–Macaulay simplicial complex. (Thus by Proposition II.4.3, every triangulation of X is Cohen–Macaulay.) Then there exists a two-dimensional contractible Cohen–Macaulay topological space X, called the *dunce hat*, such that no triangulation of X is shellable. The dunce hat is obtained by taking a solid triangle with vertices A, B, C and identifying the directed edges AB, AC, and BC. It can be embedded in $\mathbb{R}^3$ without self-intersections. One can show that X is Cohen–Macaulay and acyclic (in fact, contractible) [187]. Thus any triangulation Δ of X satisfies $h_3(\Delta) = 0$. On the other hand, every edge of Δ is contained in at least two triangles. Thus if $F_1, F_2, ..., F_s$ were a shelling of Δ, then F_s is attached to Δ_{s-1} along the entire boundary of F_s and therefore contributes 1 to h_3, a contradiction. Hence Δ is not shellable.

- It is plausible that if X is a topological space which is the geometric realization of a simplicial complex of dimension at least three, then there exists a nonshellable triangulation of X. In support of this is the fact that there exist nonshellable triangulations of the 3-sphere [115].

- The boundary complex of a simplicial polytope is shellable. This is a famous result of Bruggesser and Mani [47], [127, §5.2], [188, Thm. 8.11].

- A simplicial complex is *extendably shellable* if any partial shelling can be extended to a complete shelling. For instance, every ball or sphere of dimension two is extendably shellable [59], [83], but there exists a shellable two-dimensional simplicial complex with six vertices which is not extendably shellable [30, Exercise 7.37]. It is known that the boundary complex of a simplicial polytope need not be extendably shellable. (See [188, p. 244] for more information.) Simon [144, Ch. 5], [188, Exer. 8.24(iii)] has conjectured that the $(k-1)$-skeleton of a $(d-1)$-simplex (i.e., the simplicial complex whose facets consist of all k-element subsets of a d-element set) is extendably shellable. Though this conjecture is probably false, so far it has resisted all attempts to find a counterexample. Björner

[32, Remark 1] has extended Simon's conjecture by asking whether all matroid complexes (defined in the next section) are extendably shellable. Björner and Eriksson [32] have answered this question affirmatively for rank three matroid complexes.

We now present some material related to partitionable simplicial complexes. Since both partitionable and Cohen–Macaulay complexes have nonnegative h-vectors, it is natural to ask whether there is some connection between the two properties. We have already seen that shellable simplicial complexes are both Cohen–Macaulay and partitionable. It is easy to produce examples of partitionable complexes which are not Cohen–Macaulay, e.g., the disjoint union of the boundaries of two triangles. In fact, Björner has found an example of a partitionable complex whose f-vector is not the f-vector of a Cohen–Macaulay complex, namely, the simplicial complex with facets abc, abd, acd, bcd, def, with h-vector $(1, 3, 0, 1)$. On the other hand, a central combinatorial conjecture on Cohen–Macaulay complexes is the following.

2.7 Conjecture. *Every Cohen–Macaulay simplicial complex is partitionable.*

The next result gives a natural class of partitionable complexes which are not always shellable. (An example which isn't shellable is given by M. E. Rudin's famous example of a nonshellable triangulation of the tetrahedron [141].)

2.8 Proposition. *Let Δ be a simplicial complex which possesses a convex geometric realization (so in particular $|\Delta|$ is homeomorphic to a ball). Then Δ is partitionable.*

Proof (sketch). Let X be a convex geometric realization of Δ. Let x be a "generic" point in X. Stand at the point x and look at a facet F of Δ (by which we mean by abuse of terminology the geometric realization of F). Imagine F to be opaque and the rest of X to be transparent. Then a certain set of facets of F will be visible from x. Let G_F be the set of vertices opposite these facets. (If x is contained in F, then $G_F = \emptyset$ since F is opaque.) Then the intervals $[G_F, F]$ form a partitioning of Δ. □

The above proof is essentially the same as [111]. In this reference a more general result is actually proved, concerning the partitionability of certain triangulations of spheres. In [111] there also appears the conjecture (again in a more general context than here) that if Δ is as in

Proposition 2.8 and $\theta_1, \ldots, \theta_d$ is an l.s.o.p. for $k[\Delta]$, then the face monomials x^{G_F}, as defined in the proof of Proposition 2.8, form a k-basis for
$k[\Delta]/(\theta_1, \ldots, \theta_d)$. This conjecture is a natural extension of Theorem 2.5.
It should be pointed out that if Δ is a Cohen–Macaulay complex with
a partitioning (3), then it need not be true that the face monomials x^{G_i}
form a k-basis for $k[\Delta]$ modulo an l.s.o.p.

There are a number of other interesting results and conjectures on
various kinds of "partitionings" or "decompositions" of simplicial complexes. For instance, it is proved in [168] that every acyclic (i.e., all
reduced homology vanishes) simplicial complex Δ can be partitioned into
two-element intervals $[G, F]$ such that the set of bottom elements G form
a subcomplex of Δ. This immediately implies a result of Kalai [105]
that the f-vector of an acyclic complex is the same as the f-vector of a
cone. A far-reaching generalization was given by Duval [66]. He found
a certain kind of decomposition of Δ into one and two-element intervals
which implies a characterization of Björner and Kalai [35], [36] of those
pairs of vectors which can be simultaneously the f-vector and sequence
of Betti numbers of a simplicial complex. Several conjectures concerning
decompositions of simplicial complexes related to the acyclic case appear
in [168]. A typical one says that if both Δ and the link of every vertex
of Δ are acyclic, then Δ can be partitioned into *four*-element intervals
$[G, F]$ such that the set of bottom elements G forms a subcomplex of
Δ. The results and conjectures mentioned in this paragraph seem more
closely related to exterior algebra than commutative algebra, so we will
not say more about them here.

A recent development in the theory of shellability is its extension to
the nonpure case by Björner and Wachs [39]. We say that an arbitrary
simplicial complex Δ is *shellable* if it satisfies the condition of Definition 2.1, except that we no longer assume that Δ is pure. Björner and
Wachs show that this concept is useful for computing the homology and
homotopy type of certain nonpure simplicial complexes arising in combinatorics. We can ask whether the connections between shellability and
commutative algebra developed in this section can be extended to the
nonpure case. The central property of shellable pure simplicial complexes from the algebraic viewpoint is that they are Cohen–Macaulay.
Shellable (pure) simplicial complexes are in a sense the most tractable
class of Cohen–Macaulay complexes. Thus a basic question is to find a
"nonpure" generalization of the concept of a Cohen–Macaulay module,
so that the face ring of a shellable (nonpure) simplicial complex has this
property. It turns out that the correct definition is as follows. (We make

the definition for graded modules, but it works just as well for finitely-generated modules over local rings.)

2.9 Definition. Let M be a finitely-generated $\mathbb{Z}$-graded module over a finitely-generated connected[1] $\mathbb{N}$-graded k-algebra R. We say that M is *sequentially Cohen–Macaulay* if there exists a finite filtration

$$0 = M_0 \subset M_1 \subset \cdots \subset M_r = M \qquad (8)$$

of M by graded submodules M_i satisfying the two conditions:

(a) Each quotient M_i/M_{i-1} is Cohen–Macaulay.

(b) $\dim(M_1/M_0) < \dim(M_2/M_1) < \cdots < \dim(M_r/M_{r-1})$, where dim denotes Krull dimension.

It is easy to see that if the filtration (8) exists, then it is unique. We say that a simplicial complex Δ is *sequentially Cohen–Macaulay* (over a field k) if its face ring $k[\Delta]$ is sequentially Cohen–Macaulay. We have the following characterization of sequentially Cohen–Macaulay simplicial complexes (whose easy proof we omit).

2.10 Proposition. *Let Δ be a $(d-1)$-dimensional simplicial complex, and let Δ_i denote the subcomplex generated by the i-dimensional facets (maximal faces) of Δ. Then Δ is sequentially Cohen–Macaulay if and only if the relative simplicial complexes*

$$\Omega_i = \Delta_i/(\Delta_i \cap (\Delta_{i+1} \cup \Delta_{i+2} \cup \cdots \cup \Delta_{d-1}))$$

are Cohen–Macaulay (as defined in Section 7 of this chapter), for $0 \leq i \leq d-1$.

In particular, if $k[\Delta]$ is sequentially Cohen–Macaulay, then the subcomplex Δ_{d-1} generated by the facets of maximum dimension must be Cohen–Macaulay. Thus a pure simplicial complex is sequentially Cohen–Macaulay if and only if it is Cohen–Macaulay. It is not difficult to see (using Theorem 7.2) that sequential Cohen–Macaulayness is a *topological* property, i.e., depends only on the geometric realization $|\Delta|$ (and of course the underlying field k). One can also easily verify that shellable simplicial complexes (in the extended sense of Björner and Wachs) are sequentially Cohen–Macaulay (over any field).

---

[1] Recall from Chapter I.2 that "connected" means $R_0 = k$.

An important property of Cohen–Macaulay modules are their homological characterizations, such as Theorem I.6.3. Thus it is natural to ask whether a similar characterization exists for sequentially Cohen–Macaulay modules. Such a characterization is given by the next result. (Again, we state the result for graded modules, though everything carries over to the local case.)

2.11 Theorem. *Let M be a finitely-generated $\mathbb{Z}$-graded module over an $\mathbb{N}$-graded polynomial ring $A = k[x_1, \ldots, x_n]$ (or even over a graded Gorenstein algebra of Krull dimension n). The following two conditions are equivalent.*

(a) *M is sequentially Cohen–Macaulay.*

(b) *For all $0 \leq i \leq \dim M$, the module $\operatorname{Ext}_A^{n-i}(M, A)$ is either 0 or Cohen–Macaulay of Krull dimension i.*

The implication (a)$\Rightarrow$(b) of the preceding theorem is an immediate consequence of the long exact sequence for Ext, while the implication (b)$\Rightarrow$(a) requires a spectral sequence argument due to Christian Peskine [135].

Finally let us mention that Björner and Wachs [39, Thm. 12.3] have given an extension of Theorem 2.5 to the nonpure case.

3   Matroid complexes, level complexes, and doubly Cohen–Macaulay complexes

The following result will lead to the concept of a matroid complex.

3.1 Proposition. *Let Δ be a simplicial complex on a vertex set V. The following three conditions are equivalent.*

(a) *For every subset W of V, the induced subcomplex $\Delta_W := \{F \in \Delta : F \subseteq W\}$ is shellable.*

(b) *For every subset W of V, the induced subcomplex Δ_W is Cohen–Macaulay.*

(c) *For every subset W of V, the induced subcomplex Δ_W is pure.*

Proof. Since shellable complexes are Cohen–Macaulay and Cohen–Macaulay complexes are pure, we have (a) $\Rightarrow$ (b) $\Rightarrow$ (c). We need to show (c) $\Rightarrow$ (a).

Assume Δ satisfies (c). The proof is by induction on the number n of vertices, the assertion being clear for $n = 0$ (or $n = 1$). Now assume that Δ has the n-element vertex set V. It suffices to show that Δ is shellable, since every proper induced subcomplex satisfies (c) and hence by induction also (a). Pick a vertex $x \in V$. Let $\Delta_{V-x} = \{F \in \Delta : x \notin F\}$. If $\dim \Delta_{V-x} < \dim \Delta$ then Δ is a cone over Δ_{V-x} (since Δ is pure), i.e., $\Delta = \Delta_{V-x} \cup \{F \cup x : F \in \Delta_{V-x}\}$. (We have abbreviated $F \cup \{x\}$ as $F \cup x$.) By induction Δ_{V-x} has a shelling $F_1, \ldots, F_s$, and it is easily seen that $F_1 \cup x, \ldots, F_s \cup x$ is then a shelling of Δ.

Assume therefore that $\dim \Delta_{V-x} = \dim \Delta$. The subcomplexes Δ_{V-x} and star $x = \{F \in \Delta : F \cup x \in \Delta\}$ clearly satisfy (c), so by induction they satisfy (a). Let $F_1, \ldots, F_s$ be a shelling of Δ_{V-x} and $G_1, \ldots, G_t$ a shelling of star x. It is easy to check that $F_1, \ldots, F_s, G_1, \ldots, G_t$ is then a shelling of Δ, completing the proof. $\square$

A simplicial complex satisfying the conditions of Proposition 3.1 is called a *matroid complex*, because condition (c) is exactly the definition of "matroid" in terms of independent sets (equivalent to [132, Prop. 2.2.1]). In other words, a simplicial complex consists of the independent sets of a matroid if and only if it is a matroid complex. For further information on matroids, see for instance [184], [185], [186].

It is natural to ask whether the face ring of a matroid complex has any special algebraic properties, and whether such properties can be used to obtain combinatorial information. The property to concern us here will be a weakening of the Gorenstein property.

3.2 Proposition. *Let $R = R_0 \oplus R_1 \oplus \cdots$ be a graded Cohen–Macaulay k-algebra of Krull dimension d. Suppose that R (possibly after extending the ground field k) has an h.s.o.p. whose elements have degrees $r_1, \ldots, r_d$. Let*

$$F(R, \lambda) = \frac{h_0 + h_1\lambda + \cdots + h_s\lambda^s}{(1 - \lambda^{r_1}) \cdots (1 - \lambda^{r_d})},$$

with $h_s \neq 0$. The following twelve conditions are equivalent.

(a) *The canonical module $\Omega(R)$ of R is generated by elements all of the same degree (using the natural grading of $\Omega(R)$ defined in Chapter I.12, so that $F(\Omega(R), \lambda) = (-1)^d F(R, 1/\lambda)$).*

(b) *$\Omega(R)$ is generated by h_s elements.*

(c) *The vector space* $\mathrm{soc}(H^d(R))$ *(the socle of the local cohomology module* $H^d(R)$*) has all its elements of the same degree.*

(d) $\dim_k \mathrm{soc}(H^d(R)) = h_s$

(e) *Let* $0 \to \Lambda_{n-d} \to \cdots \to \Lambda_0 \to R \to 0$ *be a minimal free resolution of* R *as a finitely-generated module over some graded polynomial ring* A *in* n *variables. Then* Λ_{n-d} *has a basis consisting of elements all of the same degree (using the natural grading of each* Λ_i *which makes the maps in the resolution degree-preserving).*

(f) *With notation as in (e), the module* Λ_{n-d} *has rank* h_s.

(g) *With* A *as in (e), the graded vector space* $\mathrm{Tor}^A_{n-d}(R, k)$ *has all its elements of the same degree.*

(h) $\dim_k \mathrm{Tor}^A_{n-d}(R, k) = h_s$

(i) *For some h.s.o.p.* $\theta_1, \ldots, \theta_d$ *of* R, *all elements of the graded vector space* $\mathrm{soc}(R/(\theta_1, \ldots, \theta_d))$ *have the same degree.*

(j) *For some h.s.o.p.* $\theta_1, \ldots, \theta_d$ *of* R, *we have*

$$\dim_k \mathrm{soc}(R/(\theta_1, \ldots, \theta_d)) = h_s.$$

(k) *For every h.s.o.p.* $\theta_1, \ldots, \theta_d$ *of* R, *all elements of the graded vector space* $\mathrm{soc}(R/(\theta_1, \ldots, \theta_d))$ *have the same degree.*

(l) *For every h.s.o.p.* $\theta_1, \ldots, \theta_d$ *of* R, *we have*

$$\dim_k \mathrm{soc}(R/(\theta_1, \ldots, \theta_d)) = h_s.$$

Proof. These are all simple consequences of the relevant defintions and of elementary homological algebra. For instance, to relate (j) and (l) to (d), apply the long exact sequence for local cohomology (e.g., [121, Remark 3.5.3(e)]) to the short exact sequence $0 \to R \xrightarrow{\theta_1} R \to R/\theta_1 R \to 0$, and then continue to mod out by θ_2 up to θ_d. Further details are omitted. $\square$

Note that the numbers defined by conditions (b), (d), (f), (h), and (l) of Proposition 3.2 (i.e., the minimum number of generators of $\Omega(R)$, etc.) are always equal for *any* graded Cohen–Macaulay k-algebra R (or even for Cohen–Macaulay graded modules) by Theorem I.12.4; and in fact is what we defined in Chapter I.12 to be the *type* of R or M. (Theorem

I.12.4 does not explicitly mention condition (h), but this is trivially the same as (f).) Moreover, type(R) is always at least as large as h_s.

A Cohen–Macaulay ring satisfying the conditions of Proposition 3.2 is called a *level* ring. Thus R is level if and only if type(R) = h_s. For instance, Gorenstein rings are level since e.g. soc($H^d(R)$) is one-dimensional and therefore has all its elements of the same degree. The next proposition collects some facts about Hilbert functions of level rings. For more detailed proofs and other information on level rings, see [152, pp. 54–55], [40], [48, Thm. 4.3.11], [89], [90]. For a result for Cohen–Macaulay domains similar to (b) below, see [166, Thm. 2.1].

3.3 Proposition. *Let $R = R_0 \oplus R_1 \oplus \cdots$ be a standard level k-algebra, and let*
$$F(R, \lambda) = \frac{h_0 + h_1\lambda + \cdots + h_s\lambda^s}{(1 - \lambda)^d},$$
where $h_s \neq 0$.

(a) *For all i and j with $h_{i+j} > 0$ we have $h_i \leq h_j h_{i+j}$.*

(b) *If R is generically Gorenstein[2] then*
$$\sum_{i=0}^{j} h_i \leq \sum_{i=0}^{j} h_{s-i}, \quad 0 \leq j \leq s. \tag{9}$$

Proof (sketch). (a) Let $\theta_1, \ldots, \theta_d$ be an l.s.o.p. for R (extending the field k if necessary). Let $R/(\theta_1, \ldots, \theta_d) = S = S_0 \oplus \cdots \oplus S_s$, so $h_r = \dim_k S_r$. Assume $i + j \leq s$. Then the ring $S' = S_0 \oplus S_1 \oplus \cdots \oplus S_{i+j}$ is also level. The level property implies that the natural map $S_i \to \operatorname{Hom}_k(S_j, S_{i+j})$ is injective, and the proof follows.

(b) Shift the grading of the canonical module $\Omega(R)$ if necessary so that its Hilbert series is given by
$$F(\Omega(R), \lambda) = \frac{h_s + h_{s-1}\lambda + \cdots + h_0\lambda^s}{(1 - \lambda)^d}.$$

It can be shown that the property of being generically Gorenstein implies that $\Omega(R)$ contains a homogenous element u with zero annihilator, i.e., $uR \cong R$ (as R-modules). Moreover, the level property of R implies that we can take $\deg(u) = 0$.

---

[2]I.e., the localization $R_\wp$ of R at every minimal prime ideal $\wp$ is Gorenstein. Face rings in particular have this property; for them $R_\wp$ is in fact a field.

Mow an elementary algebraic argument shows that the module $M = \Omega(R)/uR$ is Cohen–Macaulay of Krull dimension $d-1$ (unless R is Gorenstein, in which case (9) is trivial since $h_i = h_{s-i}$). Thus the rational function $F(M,\lambda)(1-\lambda)^{-(d-1)}$ is in fact a polynomial with nonnegative integer coefficients. On the other hand, we have

$$
\begin{aligned}
F(M,\lambda) &= F(\Omega(R),\lambda) - F(R,\lambda) \\[2mm]
&= \frac{(h_s + h_{s-1}\lambda + \cdots + h_0\lambda^s) - (h_0 + h_1\lambda + \cdots + h_s\lambda^s)}{(1-\lambda)^d} \\[2mm]
&= \frac{\sum_{i=0}^{s-1}\left[(h_s + h_{s-1} + \cdots + h_{s-i}) - (h_0 + h_1 + \cdots + h_i)\right]\lambda^i}{(1-\lambda)^{d-1}},
\end{aligned}
$$

and the proof follows.   □

The next result gives the most significant known algebraic property of the face ring of a matroid complex. It would be interesting to discover further properties, especially if they shed some light on Conjecture 3.6 below.

3.4 Theorem. *The face ring $k[\Delta]$ of a matroid complex Δ is level.*

Proof. We claim first that if the matroid complex Δ is acyclic (i.e., has vanishing reduced homology), then Δ is a cone. This fact follows easily from some elementary matroid theory, but we will give an argument using only Proposition 3.1. Suppose Δ is acyclic but not a cone, and let $F_1, \ldots, F_s$ be a shelling of Δ (existence guaranteed by Proposition 3.1). Since Δ is acyclic, the last facet F_s contains a subfacet ($(d-2)$-dimensional face) G which is contained in no other facet of Δ. Let x be the vertex of F_s not a vertex of G. Since Δ is not a cone, there is a facet F which does not contain x. But then the induced subcomplex on the vertex set $F \cup G$ is not pure and hence not Cohen–Macaulay, thereby violating the definition of a matroid complex. This proves the claim.

It is easy to see that an arbitrary simplicial complex Γ is level if and only if the cone over Γ is level. Since removing vertices from a matroid complex keeps it a matroid complex, we may assume that Δ is not a cone. Hence by the previous paragraph Δ is not acyclic. If $\dim \Delta = d-1$, then since Δ is Cohen–Macaulay we have $\tilde{H}_{d-1}(\Delta) \neq 0$. Suppose that Δ has n vertices. Putting $i = n - d$ and $W = V$ in Theorem II.4.8 shows that the last module Λ_{n-d} in a minimal free resolution of $k[\Delta]$ over the polynomial ring $A = k[V]$ has a least one basis element of degree n. Hence we need to show that all basis elements have degree n. Suppose that W is an

$(n - j)$-element subset of V, with $j \geq 1$. By Theorem II.4.8, we need to show that $\tilde{H}_{d-j-1}(\Delta_W) = 0$. Now Δ_W is Cohen–Macaulay since Δ is a matroid complex. Hence if $\tilde{H}_{d-j-1}(\Delta_W) \neq 0$ then $\dim \Delta_W = d-j-1$. Let $x \in V - W$. Since the dimension of a simplicial complex drops by at most one when we remove a vertex, it follows that $\dim \Delta_{V-x} = \dim \Delta - 1$. But since Δ is pure, this means that Δ is a cone over Δ_{V-x}, a contradiction. $\square$

Combining Proposition 3.3 and Theorem 3.4, we see that the h-vector of a matroid complex satisfies the conditions of Proposition 3.3(a) and (b). It is natural to ask what other conditions are satisfied by the h-vector of a matroid complex. We state three such conditions in the next result. Condition (a) below is due to Brown and Colbourn [46], while (b) and (c) are due to Chari [52, Cor. 4, part 2] (who also includes a different proof of (a)) [53]. For further conditions on the h-vector (or f-vector) of a matroid complex, see [25, §2], [30, §7.5], [65].

3.5 Theorem. *Let Δ be a matroid complex with h-vector $(h_0, h_1, \ldots, h_d)$, where $h_s \neq 0$ and $h_{s+1} = \cdots = h_d = 0$.*

(a) *For any $0 \leq j \leq s$ and real number $\alpha \geq 1$, we have*

$$(-1)^j \sum_{i=0}^{j} (-\alpha)^i h_i \geq 0,$$

with equality possible only if $\alpha = 1$.

(b) $h_0 \leq h_1 \leq \cdots \leq h_{\lfloor s/2 \rfloor}$

(c) $h_i \leq h_{s-i}$ *for* $0 \leq i \leq \lfloor s/2 \rfloor$.

There is a conjecture appearing in [152, p. 59] which by a result of Hibi [90] is stronger than parts (b) and (c) of Theorem 3.5. This conjecture remains an intriguing open problem in matroid theory.

3.6 Conjecture. *Let Δ be a matroid complex with h-vector $(h_0, h_1, \ldots, h_d)$. Then there exists a pure multicomplex Γ (i.e., Γ is a multicomplex, as defined in Chapter II.2, all of whose maximal elements have the same degree) such that h_i is the number of monomials in Γ of degree i.*

A refinement of Conjecture 3.6, which we will not state here, appears in [52, Conj. 3] and gives a conjectured complete characterization (though not a numerical one like Theorems II.2.1 and II.2.2) of the h-vector of a matroid complex.

It is possible to use Theorem II.4.8 to compute the Betti numbers of $k[\Delta]$, as a module over $k[V]$, in terms of matroid-theoretic data. See [152, Thm. 9] for further details.

An interesting class of simplicial complexes related to matroid complexes and level complexes are the doubly Cohen–Macaulay (or 2-CM) complexes discussed after Theorem II.7.4. Let us expand on the discussion appearing there. First note that removing a vertex x from a pure simplicial complex Δ reduces the dimension if and only if Δ is a cone over x. It follows that a matroid complex is 2-CM if and only if it is not a cone. (In the language of matroid theory, the corresponding matroid has no *coloop* or *isthmus*.) Moreover, since every induced subcomplex of a matroid complex is Cohen–Macaulay, we see that every matroid complex is the join (defined in Chapter II.5) of a simplex and a 2-CM matroid complex.

Doubly Cohen–Macaulay complexes were first investigated by Baclawski [9], [10]. He observed that the property of being doubly Cohen–Macaulay is *local*, i.e., is preserved by links. He also showed that if L is a semimodular lattice, then the reduced order complex $\tilde{\mathcal{O}}(L)$ (defined in the paragraph preceding Proposition 4.9 of this chapter) is 2-CM if and only if L is a geometric lattice (as defined e.g. in [162, p. 105]). We mentioned in Chapter II the basic fact that the 2-CM property is topologically invariant (analogous to Proposition II.4.3 for Cohen–Macaulay complexes), as conjectured by Baclawski and proved by Jim Walker [181, Thm. 9.8]. We simply state without proof the following more precise version of Walker's result.

3.7 Proposition. *Suppose that Δ is Cohen–Macaulay (over a field k) and that $d - 1 = \dim \Delta$. Then Δ is 2-CM (over k) if and only if for all $p \in X = |\Delta|$, we have $\tilde{H}_{d-2}(X - p; k) = 0$.*

In Chapter II we also stated a homological characterization of 2-CM complexes due to Baclawski [9]; namely, a $(d - 1)$-dimensional Cohen–Macaulay complex Δ is 2-CM if and only if type $k[\Delta] = (-1)^{d-1}\tilde{\chi}(\Delta)$. The proof is a fairly straightforward application of Theorem II.4.8 and is repeated in [48, Thm. 5.6.6]. Since for any $(d - 1)$-dimensional simplicial complex Δ we have $(-1)^{d-1}\tilde{\chi}(\Delta) = h_d$ (see observation (1) at the end of Chapter II.2), it follows from the definition of a level ring that the face ring of a 2-CM complex is level. Thus a level complex Δ is 2-CM if and only if the integer s of Proposition 3.2 is equal to d. As an immediate consequence of Baclawski's homological characterization of 2-CM, we see that a 2-CM complex Δ cannot be acyclic, since a Cohen–Macaulay complex Δ is acyclic if and only if $\tilde{\chi}(\Delta) = 0$, and since $|\tilde{\chi}(\Delta)| = \text{type}(k[\Delta]) > 0$ for 2-CM complexes.

4 Balanced complexes, order complexes, and flag complexes

The underlying idea in this section is that for certain simplicial complexes we can refine the $\mathbb{N}$-grading of the face ring and still have a *homogeneous* system of parameters.

4.1 Definition. Let Δ be a $(d-1)$-dimensional simplicial complex on the vertex set V. We say that Δ is *balanced* if there is a map $\kappa : V \to [d]$ such that if $\{x, y\}$ is an edge of Δ, then $\kappa(x) \neq \kappa(y)$. Equivalently, if we think of κ as a *coloring* of V with the colors $1, 2, \ldots, d$, then every face of Δ has all its vertices colored differently. We will always assume that the coloring κ is part of the structure of a balanced complex, even if it is not mentioned explicitly.

NOTE. In the original definition [156, §2] of a balanced complex it was also assumed that Δ is pure, but there is no real reason for this restriction. We will be concerned exclusively here with Cohen–Macaulay balanced complexes, which are always pure by Reisner's theorem (Corollary II.4.2). However, Theorem 4.6 shows that nonpure balanced complexes arise naturally in the theory of Cohen–Macaulay balanced complexes. Also in [156] the term "completely balanced" was used for our "balanced" here. There was a more general notion of a (partially) balanced complex. Since the most interesting examples of partially balanced complexes are completely balanced, we will be concerned here only with the completely balanced case.

Balanced complexes have a natural refinement of the f-vector. If Δ is balanced of dimension $d - 1$ and if S is a subset of $[d]$, then define $\alpha_\Delta(S)$ (or just $\alpha(S)$) to be the number of faces F of Δ whose vertex colors are precisely the elements of S, i.e., $\kappa(F) = S$. Thus the usual f-vector $f(\Delta) = (f_0, \ldots, f_{d-1})$ is given by

$$f_i(\Delta) = \sum_{|S|=i+1} \alpha_\Delta(S).$$

The function α_Δ is called the *flag f-vector* of Δ. There is also a natural analogue of the h-vector. For $S \subseteq [d]$, define

$$\beta(S) = \beta_\Delta(S) = \sum_{T \subseteq S} (-1)^{|S-T|} \alpha_\Delta(T). \tag{10}$$

Equivalently (by the Principle of Inclusion-Exclusion [162, Thm. 2.1.1]),

$$\alpha(S) = \sum_{T \subseteq S} \beta(T). \tag{11}$$

We can also express the relationship between α and β in terms of generating functions, analogous to equation (4). Let $\lambda_1, \ldots, \lambda_d$ be indeterminates. Then it follows immediately from the relationship (10) between α and β that

$$\sum_S \alpha(S) \left(\prod_{i \in S} \lambda_i \right) \left(\prod_{i \in [d]-S} (1 - \lambda_i) \right) = \sum_S \beta(S) \prod_{i \in S} \lambda_i. \tag{12}$$

If we set each $\lambda_i = 1/x$ and multiply by x^d in (12), then comparing with (4) shows that

$$h_i(\Delta) = \sum_{|S|=i} \beta_\Delta(S).$$

Hence β_Δ is a "natural" refinement of the h-vector of Δ, and we call the function β_Δ the *flag h-vector* of Δ.

Unlike the situation with the ordinary h-vector, the flag h-vector has a simple topological interpretation. For $S \subseteq [d]$ define the *S-rank-selected subcomplex* Δ_S of Δ by

$$\Delta_S = \{F \in \Delta : \kappa(F) \subseteq S\}.$$

It is easy to see that

$$\beta_\Delta(S) = (-1)^{|S|-1} \tilde{\chi}(\Delta_S), \tag{13}$$

where $\tilde{\chi}(\Delta_S)$ denotes the reduced Euler characteristic of Δ_S [156, Prop. 3.5].

Let us now consider the face ring $k[\Delta]$ of a balanced complex Δ with coloring κ. Let e_i denote the ith unit coordinate vector in $\mathbb{N}^d$, and for a vertex x of Δ define $\deg x = e_{\kappa(x)} \in \mathbb{N}^d$. This definition makes $k[\Delta]$ into an $\mathbb{N}^d$-graded k-algebra. Thus the Hilbert series $F(k[\Delta], \lambda)$ (abbreviated $F(\Delta, \lambda)$) in the variables $\lambda = (\lambda_1, \ldots, \lambda_d)$ is given by

$$F(\Delta, \lambda) = \sum_{\gamma \in \mathbb{N}^d} \dim_k(k[\Delta]_\gamma) \lambda^\gamma.$$

4.2 Proposition. *We have*

$$F(\Delta, \lambda) = \frac{\sum_S \beta_\Delta(S)\lambda^S}{(1 - \lambda_1) \cdots (1 - \lambda_d)},$$

where $\lambda^S = \prod_{i \in S} \lambda_i$.

Proof. It is clear that

$$F(\Delta, \lambda) = \sum_S \alpha_\Delta(S) \prod_{i \in S} \frac{\lambda_i}{1 - \lambda_i}.$$

Put each term over the common denominator $(1 - \lambda_1) \cdots (1 - \lambda_d)$, and comparing with (12) completes the proof. □

Another important property of the face ring of a balanced complex is the following. We will see (Theorems 4.4–4.6) that it is an example of the importance of choosing a "special" system of parameters.

4.3 Proposition. (a) *Let* Δ *be a balanced simplicial complex of dimension* $d - 1$ *on the vertex set* V. *Define*

$$\theta_i = \sum_{\substack{x \in V \\ \kappa(x) = i}} x \in k[\Delta],$$

$1 \le i \le d$. *(Note that* θ_i *is homogeneous of degree* $e_i \in \mathbb{N}^d$.*) Then* $\theta_1, \ldots, \theta_d$ *is an h.s.o.p. for* $k[\Delta]$.

(b) *Let* $S = k[\Delta]/(\theta_1, \ldots, \theta_d)$. *Then for every* $x \in V$, *we have* $x^2 = 0$ *in* S.

Proof. It suffices to prove (b), since then S is spanned by squarefree monomials and hence is a finite-dimensional vector space. (On the other hand, it is easy to prove (a) directly from Lemma 2.4.) Let $x \in V$, and suppose $\kappa(x) = i$. If $y \in V$, $y \ne x$, and $\kappa(y) = i$, then $\{x, y\}$ is not an edge of Δ so $xy = 0$ in $k[\Delta]$. Hence $x\theta_i = x^2$ in $k[\Delta]$, so $x^2 = 0$ in S. □

Note that in Proposition 4.3(a), we can take θ_i to be *any* linear combination of vertices colored i provided no coefficient is zero. Let us call the h.s.o.p. of Proposition 4.3 the *canonical* h.s.o.p. for $k[\Delta]$.

We come to three basic properties of balanced Cohen–Macaulay complexes.

4.4 Theorem. *Let Δ be a $(d-1)$-dimensional balanced Cohen–Macaulay complex. Let $\theta_1, \ldots, \theta_d$ be the canonical h.s.o.p. for $k[\Delta]$, and let $S = k[\Delta]/(\theta_1, \ldots, \theta_d)$. Then the Hilbert series of S (as an $\mathbb{N}^d$-graded algebra) is given by*

$$F(S, \lambda) = \sum_{T \subseteq [d]} \beta_\Delta(T) \lambda^T.$$

Hence $\beta_\Delta(T) \geq 0$ for all T.

Proof. By the characterization of homogeneous regular sequences mentioned after Definition I.5.6, we have

$$
\begin{aligned}
F(S, \lambda) &= F(\Delta, \lambda) \prod_{i=1}^{d} \left(1 - \lambda^{\deg \theta_i} \right) \\
&= F(\Delta, \lambda) \prod_{i=1}^{d} (1 - \lambda_i).
\end{aligned}
$$

Now use Proposition 4.2. $\qquad\qquad\qquad\qquad\qquad\qquad\qquad\qquad\square$

4.5 Theorem. *Let Δ be a $(d-1)$-dimensional balanced Cohen–Macaulay complex. Then for every $S \subseteq [d]$, the rank-selected subcomplex Δ_S is Cohen–Macaulay.*

Proof. A purely topological proof can be given using Corollary II.4.2 [181, Thm. 9.4], [31, (11.13)]. We will give instead a simple proof based on commutative algebra. Namely, since $k[\Delta]$ is Cohen–Macaulay we have by Theorem I.5.10 a decomposition

$$k[\Delta] = \coprod_{\eta \in X} \eta \cdot k[\theta], \tag{14}$$

where X is some set of monomials in the vertices of Δ, and $\theta = (\theta_1, \ldots, \theta_d)$ is the canonical h.s.o.p. for $k[\Delta]$. Let I_S be the ideal of $k[\Delta]$ generated by all vertices x satisfying $\kappa(x) \notin S$. Let X_S denote the subset of monomials in X all of whose variables are colored from S, and let θ_S denote the subsequence $(\theta_i : i \in S)$ of θ. It is then clear from (14) that

$$k[\Delta]/I_S = \coprod_{\eta \in X_S} \eta \cdot k[\theta_S].$$

But $k[\Delta_S] \cong k[\Delta]/I_S$, so $k[\Delta_S]$ is a free module over a system of parameters. Hence $k[\Delta_S]$ is Cohen–Macaulay. $\qquad\qquad\qquad\square$

Note that Theorem 4.5, together with equation (13), gives another proof that $\beta_\Delta(S) \geq 0$ for balanced Cohen–Macaulay complexes Δ, since for any $(m - 1)$-dimensional Cohen–Macaulay complex Γ we have

$$(-1)^{m-1}\tilde{\chi}(\Gamma) = h_m(\Gamma) \geq 0 \ .$$

4.6 Theorem. *Let $\beta : 2^{[d]} \to \mathbb{Z}$. The following two conditions are equivalent.*

(a) *There exists a balanced $(d - 1)$-dimensional Cohen–Macaulay complex Δ such that $\beta_\Delta = \beta$.*

(b) *There exists a balanced simplicial complex (which we are assuming need not be pure) Λ such that $\alpha_\Lambda(S) = \beta_\Delta(S)$ for all $S \subseteq [d]$.*

Proof. (a) $\Rightarrow$ (b). Let $\theta_1, \ldots, \theta_d$ be the canonical h.s.o.p. for $k[\Delta]$, and let $C = k[\Delta]/(\theta)$. Just as in the sketch of the proof of Theorem II.2.3, choose the lexicographically earliest basis of monomials for S. These monomials will form a multicomplex Λ. By Proposition 4.3(b), Λ is in fact a simplicial complex. Color the vertices of Λ by the same colors as in Δ. By Theorem 4.4, every face of Λ has all its vertices colored differently (in fact, Λ is isomorphic to a subcomplex, though not necessarily induced, of Δ), and $\alpha_\Lambda(S) = \beta_\Delta(S)$. Hence (a) $\Rightarrow$ (b).

(b) $\Rightarrow$ (a). This is a construction given in [33]. (It is stated on page 152 of [156] that the implication (b) $\Rightarrow$ (a) is *false*, but this statement is erroneous.) $\qquad \square$

Note that Theorem 4.6 gives a characterization of the flag h-vector (or equivalently, the flag f-vector) of a balanced Cohen–Macaulay complex analogous to the characterization (implicit in Chapter II.2) of the h-vector of a Cohen–Macaulay complex as the f-vector of a multicomplex. A purely numerical characterization of the flag f-vector of a balanced complex, analogous to the Kruskal-Katona theorem (Theorem II.2.1), appears in [76].

An important class of balanced complexes is defined as follows. Let P be a finite partially ordered set (or *poset*, for short), and let $\mathcal{O}(P)$ be the set of chains of P. (For basic information on finite posets, see [162, Ch. 3].) Then $\mathcal{O}(P)$ is a simplicial complex, called the *order complex* of P. If we color an element x of P by the cardinality i of the longest chain $x_1 < x_2 < \cdots < x_i = x$ of P with top element x, then $\mathcal{O}(P)$ becomes a balanced simplicial complex. The statement that P has a certain property we will often take to mean that $\mathcal{O}(P)$ has this property. In particular we have the notion of a *Cohen–Macaulay poset*. Moreover, we write $k[P]$ as short for $k[\mathcal{O}(P)]$, the face ring of the order complex of P. Some salient properties of Cohen–Macaulay posets will be discussed below. For further information, see [7] and [34].

A basic property of order complexes is that every minimal nonface has two elements, i.e., the ideal I_Δ of Definition II.1.1 is generated by elements of degree two. (For a complete characterization of order complexes, see [156, p. 144].) General simplicial complexes for which every minimal nonface has two elements are called *flag complexes* (after Tits [179, p. 2]). Flag complexes are a fascinating class of simplicial complexes which deserve further study. To support this statement, we mention four results or conjectures related to flag complexes.

- Much deep work related to flag complexes has been done by graph theorists, in the guise of "Turán-type" theorems; see e.g., [116, 10.30–10.40].

- Let $(f_0, f_1, \ldots, f_{d-1})$ be the f-vector of a flag complex. It follows from [180, Prop. 5.10] (see also [173, Cor. 2.4]) that if we define

$$\log\left(1 - f_0 t + f_1 t^2 - \cdots + (-1)^d f_{d-1} t^d\right) = -\sum_{n \geq 1} b_n \frac{x^n}{n},$$

  then each b_n is a *nonnegative integer*. (Integrality is easy; nonnegativity is the interesting fact.)

- Kalai has conjectured (unpublished) that the f-vector of a flag complex Δ is also the f-vector of a balanced complex Γ. Moreover, if Δ is Cohen–Macaulay then we can take Γ to be Cohen–Macaulay. Hence by Theorem 4.6 it would follow that the h-vector of a Cohen–Macaulay flag complex is the f-vector of a balanced complex.

- Let Δ be a $(d-1)$-dimensional nonacyclic *Gorenstein* flag complex. Let $h(\Delta) = (h_0, \ldots, h_d)$, and suppose that d is even, say $d = 2e$. Then Charney and Davis conjecture [54, §4], [55, Conjecture D], as a discrete analogue of a well-known conjecture of Hopf on the Euler characteristic of a nonpositively curved, piecewise Euclidean manifold, that

$$(-1)^e (h_0 - h_1 + h_2 - \cdots + h_d) \geq 0.$$

This conjecture is open even for nonacyclic Gorenstein order complexes. We will say more about this special case below.

An important feature of order complexes is the existence of a simple combinatorial technique, called "lexicographic shellability," which yields when applicable an explicit shelling of the order complex $\mathcal{O}(P)$, thereby showing that P is Cohen–Macaulay. Lexicographic shellability can be useful even for simplicial complexes which are not order complexes, since a simplicial complex Δ is Cohen–Macaulay if and only if its face poset P_Δ (set of faces, partially ordered by inclusion) is Cohen–Macaulay. This is because the order complex $\mathcal{O}(P_\Delta)$ of P_Δ is just the first barycentric subdivision of Δ, so that Δ and P_Δ have homeomorphic geometric realizations. Hence Δ is Cohen–Macaulay if and only if P_Δ is Cohen–Macaulay, by the topologicalness of the Cohen–Macaulay property (Proposition II.4.3). For the definition, properties, and applications of lexicographic shellability, see for instance [26], [34, §2], [37], [38]. An extension to nonpure posets appears in [39], while an extension to relative posets (using the terminology of Section 7) appears in [172, §4].

It is natural to ask whether the characterization of flag f-vectors of Cohen–Macaulay balanced complexes given by Theorem 4.6 can be strengthened in the case of Cohen–Macaulay posets. It seems very difficult to obtain a complete characterization of the flag f-vector of a Cohen–Macaulay poset. There is a special case for which some especially interesting results and conjectures exist. This is the case of Cohen–Macaulay posets which are also nonacyclic and Gorenstein. (The nonacyclicity is a harmless assumption, because of the equivalence of (a) and (c) in Theorem II.5.1.) It is convenient to deal with posets which possess a unique minimal element $\hat{0}$ and unique maximal element $\hat{1}$. Thus define an *Eulerian poset* to be a graded poset P (i.e., every maximal chain has the same length) with $\hat{0}$ and $\hat{1}$ whose Möbius function μ (as defined, e.g., in [162, Ch. 3.7]) satisfies $\mu(s,t) = (-1)^{\ell(s,t)}$ for all $s \leq t$ in P, where $\ell(s,t)$ denotes the length of the closed interval $[s,t] := \{u \in P : s \leq u \leq t\}$. Equivalently, an Eulerian poset is a graded poset with $\hat{0}$ and $\hat{1}$ such that every interval with more than one element has the same number of elements of even rank as of odd rank. It is an immediate consequence of Theorem II.5.1 and the theorem of Philip Hall [162, Prop. 3.8.5] relating Möbius functions and Euler characteristics that a Cohen–Macaulay poset Q is nonacyclic and Gorenstein if and only if the poset $\hat{Q} = Q \cup \{\hat{0}, \hat{1}\}$ is Eulerian. For further information on Eulerian posets, see [162, Ch. 3.14] and [170].

If P is Eulerian of rank $n + 1$, then by slight abuse of notation we let α_P and β_P denote the flag f-vector and flag h-vector of $P - \{\hat{0}, \hat{1}\}$. We will define certain noncommutative generating functions for α_P and

β_P. Let a and b be noncommuting variables, and for $S \subseteq [n]$ define a noncommutative monomial $u_S = u_1 u_2 \cdots u_n$ by

$$u_i = \begin{cases} a, & \text{if } i \notin S \\ b, & \text{if } i \in S. \end{cases}$$

Define the generating functions

$$\Upsilon_P(a,b) = \sum_{S \subseteq [n]} \alpha_P(S) u_S$$

$$\Psi_P(a,b) = \sum_{S \subseteq [n]} \beta_P(S) u_S.$$

These definitions can be made for any graded poset, and it is clear that equations (10) and (11) are equivalent to the relations

$$\Psi_P(a,b) = \Upsilon_P(a, b - a)$$

$$\Upsilon_P(a,b) = \Psi_P(a, a + b).$$

In a seminal paper [12] on flag vectors of Eulerian posets, M. Bayer and L. Billera found the most general linear equalities satisfied by such vectors. This result is the analogue, for flag vectors of Eulerian posets, of the Dehn-Sommerville equations (Chapter II.6) for h-vectors of Gorenstein simplicial complexes (or more generally, simplicial complexes such that the link of every face has the Euler characteristic of a sphere of its dimension). J. Fine realized that the Bayer-Billera result could be stated as follows. (See [74] for related results.)

4.7 Theorem. *Let P an Eulerian poset. Then there exists a polynomial $\Phi_P(c, d)$ in the noncommuting variables c and d such that*

$$\Psi_P(a,b) = \Phi_P(a + b, ab + ba). \qquad \square$$

The polynomial $\Phi_P(c,d)$ is called the *cd-index* of P. For further information and references on the *cd*-index, see [171]. Here we wish simply to point out a connection between commutative algebra, the *cd*-index, and the Charney-Davis conjecture. Perhaps the most intriguing open problem concerning the *cd*-index is the following [171, Conj. 2.1].

4.8 Conjecture. *Suppose that P is Eulerian and Cohen–Macaulay. Then $\Phi_P(c, d) \geq 0$, i.e., every coefficient of $\Phi_P(c, d)$ is nonnegative.*

The main significance of Conjecture 4.8 is that if it were true then it would imply all linear *inequalities* satisfied by the flag f-vector (or flag h-vector) of a Cohen–Macaulay Eulerian poset [171, Thm. 2.1]. Eric Babson observed that if P is Eulerian of rank $2m+1$, then the coefficient of d^m in $\Phi_P(c, d)$ is equal to $(-1)^m(h_0 - h_1 + h_2 - \cdots + h_{2m})$, where $(h_0, h_1, \ldots, h_{2m})$ denotes the h-vector of the order complex $\mathcal{O}(P)$ (or of $\mathcal{O}(P - \{\hat{0}, \hat{1}\})$). It follows that Conjecture 4.8 implies the Charney-Davis conjecture for the special case of order complexes. For certain Eulerian posets which include face lattices of convex polytopes an inductive proof of Conjecture 4.8 can be given [171, Thm. 2.2], and hence the Charney-Davis conjecture is true for these posets.

We now discuss the connection between the Charney-Davis conjecture for order complexes and commutative algebra, based on an idea of Gil Kalai. This is the only known algebraic interpretation (though only conjecturally) of a nontrivial cd-index coefficient. It would be very interesting to be able to relate the entire cd-index, say in the Cohen–Macaulay case, to algebraic objects like Cohen–Macaulay rings or toric varieties. Let $P = P_0 \cup P_1 \cup \cdots \cup P_{2m+1}$ (disjoint union) be a Cohen–Macaulay Eulerian poset of rank $2m + 1$, where P_i denotes the set of elements of rank i. Thus by Theorem II.5.1 P is Gorenstein. In fact, $P - \{\hat{0}, \hat{1}\}$ is nonacyclic Gorenstein of dimension $2m - 1$, so the Charney-Davis conjecture applies to the "reduced order complex" $\tilde{\mathcal{O}}(P) = \mathcal{O}(P - \{\hat{0}, \hat{1}\})$. Let k be a "large" field *of characteristic two.* ("Large" means that k has sufficiently many independent transcendental elements over its prime field $\mathbb{F}_2$.) Let

$$\theta_i = \sum_{x \in P_i} x \in k[P], \quad 0 \le i \le 2m + 1.$$

By Proposition 4.3(a), $\theta_0, \ldots, \theta_{2m+1}$ is an l.s.o.p. for $k[P]$. Let

$$R = k[P]/(\theta_0, \ldots, \theta_{2m+1}).$$

Thus the Hilbert function of R is given by $H(R, i) = h_i(P)$ (short for $h_i(\mathcal{O}(P))$). Now let ω be a generic linear form in R, i.e., $\omega = \sum_{x \in P} c_x x$, where the c_x's are independent transcendentals in k. We are in a situation somewhat like that of Theorem 1.4. If k had characteristic zero it would be in fact reasonable to expect (though we don't know whether it's actually true) that the multiplication $\omega : R_i \to R_{i+1}$ is *injective* for $0 \le i < m$. This cannot be the case, however, in the present situation, since Proposition 4.3(b) and the fact that $\text{char}(k) = 2$ shows that $\omega^2 = 0$. Hence (writing ω_j for ω restricted to R_j, so that ω_j is a linear transformation $\omega_j : R_j \to R_{j+1}$) we have that $\ker(\omega_i) \subseteq \operatorname{im}(\omega_{i-1})$. Thus the "most injective" that ω_i could possibly be for $i < m$ is that $\ker(\omega_i) = \operatorname{im}(\omega_{i-1})$.

4.9 Proposition. *Suppose that $\ker(\omega_i) = \operatorname{im}(\omega_{i-1})$ for all $i < m$. Then $(-1)^m \sum_{i=0}^{2m} (-1)^i h_i \geq 0$, i.e., the Charney-Davis conjecture holds for the order complex of $P - \{\hat{0}, \hat{1}\}$.*

Proof. Since R is Gorenstein with socle in degree $2m$, the pairing $R_i \times R_{2m-i} \to R_{2m}$ is a perfect pairing, i.e., the natural map $R_i \to \operatorname{Hom}_k(R_{2m-i}, R_{2m})$ is a bijection. From this it is easy to deduce that $\ker(\omega_i) = \operatorname{im}(\omega_{i-1})$ also for $i > m$. Consider the diagram

$$\mathcal{K} : 0 \to R_0 \xrightarrow{\omega} R_1 \xrightarrow{\omega} \cdots \xrightarrow{\omega} R_{2m} \to 0.$$

$\mathcal{K}$ is a complex (over k) since $\omega^2 = 0$. Since $\ker(\omega_i) = \operatorname{im}(\omega_{i-1})$ for $i \neq m$, all the homology of this complex is concentrated in dimension m. Hence if $\chi(\mathcal{K})$ denotes the Euler characteristic of $\mathcal{K}$, then

$$
\begin{aligned}
(-1)^m \chi(\mathcal{K}) &= (-1)^m \sum_{i=0}^{2m} (-1)^i \dim_k H_i(\mathcal{K}) \\
&= \dim_k H_m(\mathcal{K}) \\
&\geq 0.
\end{aligned}
$$

But also

$$
\begin{aligned}
(-1)^m \chi(\mathcal{K}) &= (-1)^m \sum_{i=0}^{m} (-1)^i \dim_k R_i \\
&= (-1)^m \left(h_0 - h_1 + h_2 - \cdots + h_{2m} \right),
\end{aligned}
$$

and the proof follows. $\square$

We suspect that the hypothesis in Proposition 4.9 that $\ker(\omega_i) = \operatorname{im}(\omega_{i-1})$ for $i < m$ will be difficult to prove since it is a kind of hard Lefschetz theorem, but without any associated variety. Nevertheless, Proposition 4.9 adds to the plausibility of the Charney-Davis conjecture, at least for order complexes.

It is reasonable to ask whether there are interesting examples of balanced complexes other than order complexes of posets. One such class of examples are the Coxeter complexes and Tits buildings, which we will briefly discuss here. Much of our presentation will follow Björner [28]. A finite *Coxeter group* (W, S) may be thought of as a finite subgroup W of the orthogonal group $O(\mathbb{R}^{|S|})$ generated by reflections, where S is the set of reflections through hyperplanes which bound some fixed connected region in the complement of the union of all reflecting hyperplanes. The set S is then a minimal generating set for W, and W has a simple presentation in terms of S. For a panoply of equivalent definitions, as well as much additional information, see [43], [99]. Subgroups W_J generated by subsets $J \subseteq S$ are called *parabolic*. In particular, a *maximal* parabolic subgroup W_J satisfies $|S - J| = 1$. The archetypal example of a finite Coxeter group is the symmetric group $\mathfrak{S}_n$ of all permutations of the set $\{1, 2, \ldots, n\}$, with the set S consisting of the adjacent transpositions $(i, i + 1)$, $1 \le i \le n - 1$. There is a well-known classification of finite Coxeter groups as direct products of elements of the infinite families A_n $(n \ge 1)$, B_n $(n \ge 2)$, D_n $(n \ge 4)$, $I_2(p)$ $(p = 5$ or $p \ge 7)$, and the exceptional groups E_6, E_7, E_8, F_4, G_2, H_3, and H_4. See [43, Ch. VI, §4.1, Thm. 1, pp. 193–194].

Let (W, S) be a finite Coxeter group. The *Coxeter complex* $\Delta(W, S)$ is by definition the nerve of the covering of W by left cosets of maximal parabolic subgroups of W. In other words, the vertices of (W, S) are the left cosets of maximal parabolic subgroups, and a set F of vertices forms a face if the intersection of the elements of F is a nonempty subset of W. Thus $\Delta(W, S)$ is a simplicial complex. The maximal faces (or *chambers*) are in one-to-one correspondence with the elements of W. More precisely, the chambers are the sets

$$C_w = \{wW_{S-s} : s \in S\}, \quad w \in W.$$

The Coxeter complex is an abstract reformulation of the simplicial decomposition of the unit sphere in $\mathbb{R}^{|S|}$ which is produced by the intersections with reflecting hyperplanes belonging to W. Its abstract formulation and a characterization in these terms is due to Tits [179].

Since $\Delta(W, S)$ is a simplicial decomposition of a sphere, it is Cohen–Macaulay by Corollary II.4.4. Moreover, if we color the vertex wW_{S-s} with the "color" s, then $\Delta(W, S)$ becomes a balanced complex. An irreducible Coxeter complex (i.e., a Coxeter complex corresponding to a Coxeter group that is not a direct product of smaller Coxeter groups) is the order complex of a poset exactly for the types A_n, B_n, F_4, G_2, H_3, H_4, and $I_2(p)$ (whose Coxeter diagrams contain no "forks," i.e., vertices incident to three other vertices). Thus the Coxeter complexes of types D_n, E_6, E_7, and E_8 are "natural" balanced complexes that are not order complexes. The usual combinatorial data associated with balanced Cohen–Macaulay complexes has an elegant group-theoretical meaning for

Coxeter complexes. In particular, if $J \subseteq S$, then the flag h-vector of $\Delta(W, S)$ is given by

$$\beta_{\Delta(W,S)}(J) = \#\{w \in W : \ell(ws) < \ell(w) \text{ if and only if } s \in J\}, \qquad (15)$$

where ℓ is the length function of W, i.e., $\ell(w)$ is the smallest integer m such that w can be written as a product of m elements of S. For further details, see [27], [28].

A *building* is a kind of "q-analogue" of a Coxeter group, due to Tits [179]. Buildings are also balanced Cohen–Macaulay complexes. Rather than giving the general definition here, let us just consider one example. Let $B_n(q)$ denote the lattice of subspaces of the vector space $\mathbb{F}_q^n$, ordered by inclusion. This lattice is a well-known q-analogue of the boolean algebra B_n of all subsets of $\{1, 2, \ldots, n\}$. This means that many formulas involving $B_n(q)$ reduce to the corresponding formulas for B_n when we set $q = 1$. For instance, $B_n(q)$ has $(1 + q)(1 + q + q^2) \cdots (1 + q + \cdots + q^{n-1})$ maximal chains. When we set $q = 1$ we obtain $n!$, the number of maximal chains of B_n. The reduced order complex $\tilde{\mathcal{O}}(B_n)$ is easily seen to be isomorphic to the Coxeter complex of type A_{n-1}, so $\tilde{\mathcal{O}}(B_n(q))$ may be regarded as the q-analogue of this order complex; and in fact $\tilde{\mathcal{O}}(B_n(q))$ is isomorphic to the building of type A_{n-1} (over the field $\mathbb{F}_q$). Let J be a subset of $\{1, 2, \ldots, n - 1\}$. The flag h-vector of $\tilde{\mathcal{O}}(B_n(q))$ is given by

$$\beta_{\tilde{\mathcal{O}}(B_n(q))}(J) = \sum q^{\ell(w)}, \qquad (16)$$

summed over all $w \in \mathfrak{S}_n$ satisfying $\ell(ws) < \ell(w)$ if and only if $s \in J$. Note that equation (16) reduces to (15) (in the case $W = \mathfrak{S}_n$) when $q = 1$. For further information on buildings, see for instance [27], [45], [179].

5   Splines

There is a remarkable connection between face rings and shellability on the one hand, and the theory of splines on the other. Spline theory is of fundamental importance in numerical analysis and computer aided design. The use of algebraic techniques to investigate splines was pioneered by Louis Billera, partially in collaboration with Lauren Rose [17], [18], [22], [23], [24]. We will point out one aspect of the algebraic theory of splines that is closely related to face rings. Our treatment will follow [18].

Let Δ be a pure m-dimensional simplicial complex, rectilinearly embedded in $\mathbb{R}^m$. We will think of Δ as a subset of $\mathbb{R}^m$ with the additional

structure of an embedded simplicial complex. Define $C^r(\Delta)$ to be the set
of all functions $f : \Delta \to \mathbb{R}$ such that (a) all the rth order partial deriva-
tives of f exist and are continuous, and (b) the restriction $f|_F$ of f to
any face (or equivalently, any facet) of Δ is given by a (real) polynomial
in m variables (the coordinates of $\mathbb{R}^m$). The elements of $C^r(\Delta)$ are called
piecewise polynomials, splines, or *finite elements.*

The set $C^r(\Delta)$ forms a vector space over $\mathbb{R}$, and has a natural filtration

$$C_0^r(\Delta) \subseteq C_1^r(\Delta) \subseteq \cdots,$$

where $C_j^r(\Delta)$ consists of all functions f such that each polynomial f_F is
of degree at most j. Strang [174], [175] was the first to raise explicitly
the question of finding the dimension and a basis for the spaces $C_j^r(\Delta)$.
Here we will be concerned with the case $r = 0$, i.e., f is required only
to be *continuous* and piecewise polynomial. (Continuity is actually an
immediate consequence of the piecewise polyomial property.) For further
discussion of the general case, see [17], [22], [23], [24].

The vector spaces $C^r(\Delta)$ have additional algebraic structure. A multivariate form of the chain rule shows that $C^r(\Delta)$ is a ring under pointwise
multiplication. Moreover, the restriction to Δ of any globally polynomial
function on $\mathbb{R}^m$ is an element of $C^r(\Delta)$. Hence $C^r(\Delta)$ contains a subalgebra $R = \mathbb{R}[y_1, \ldots, y_m]$, isomorphic to a polynomial ring in m variables,
where y_i is just the projection onto the ith coordinate in $\mathbb{R}^m$. Thus in
particular $C^r(\Delta)$ is a module over the polynomial ring R.

From now on we deal only with the case $r = 0$. We will for the most
part simply state without proof the basic results from [18]. We first want
to describe a set of generators for $C^0(\Delta)$ as an $\mathbb{R}$-algebra. Suppose that
the vertices of Δ are $v_1, \ldots, v_n$. Let X_i be the unique piecewise linear
function on Δ defined by $X_i(v_j) = \delta_{ij}$, where δ_{ij} is the Kronecker delta.
It is straightforward to see that $X_1, \ldots, X_n$ is a basis for the real vector
space $C_1^0(\Delta)$. Consideration of this basis goes back to Courant [56], and
the functions X_i are called the *Courant functions* of Δ. The first main
result of Billera on the algebra $C^0(\Delta)$ is the following [18, Thm. 2.3]
(proved by induction on the number of facets).

5.1 Theorem. *For any d-dimensional simplicial complex Δ, the space
$C^0(\Delta)$ is generated as an $\mathbb{R}$-algebra by its Courant functions $X_1, \ldots, X_n$.*

The next step is to find the relations satisfied by the generators
$X_1, \ldots, X_n$. It is easily seen from the definition of the Courant func-
tions that if $\{v_{i_1}, v_{i_2} \ldots, v_{i_p}\}$ is a set of vertices of Δ that do not form a

face, then
$$X_{i_1} X_{i_2} \cdots X_{i_p} = 0.$$

Hence there is a natural surjection $\mathbb{R}[\Delta] \to C^0(\Delta)$, where $\mathbb{R}[\Delta]$ is the face ring of Δ (regarded as an abstract simplicial complex) in the variables $x_1, \ldots, x_n$, sending x_i to X_i.

It is also immediate from the definition of the Courant functions that they satisfy the additional relation that $X_1 + X_2 + \cdots + X_n$ is the constant function 1. Hence the polynomial $x_1 + x_2 + \cdots + x_n - 1$ belongs to the kernel of the homomorphism $\mathbb{R}[\Delta] \to C^0(\Delta)$, so we have a surjective algebra homomorphism $\varphi : \mathbb{R}[\Delta]/(x_1 + \cdots + x_n - 1) \to C^0(\Delta)$. Billera's second main theorem [18, Thm. 3.6] (also proved by induction on the number of facets) is the following.

5.2 Theorem. *The homomorphism φ defined above is an isomorphism of $\mathbb{R}$-algebras.*

As an immediate application of Theorem 5.2, we can compute the dimension of the vector spaces $C_j^0(\Delta)$.

5.3 Lemma. *The canonical homomorphism*

$$\psi : \mathbb{R}[\Delta] \to \mathbb{R}[\Delta]/(x_1 + \cdots + x_n - 1) := S$$

takes the vector space $\mathbb{R}[\Delta]_j$ bijectively onto the space S_j spanned by all polynomials in the variables $x_1, \ldots, x_n$ of degree at most j.

Proof. Let M be the set of all monomials of degree j whose support is a face of Δ. Hence M is a basis for $\mathbb{R}[\Delta]_j$. We claim that the images of these monomials are linearly independent in S. Otherwise we have a relation in $\mathbb{R}[\Delta]$ of the form

$$\sum_{u \in M} \alpha_u u = \left(\sum x_i - 1 \right) g, \tag{17}$$

where $\alpha_u \in \mathbb{R}$ (not all zero), and $0 \neq g \in \mathbb{R}[\Delta]$. Now if v is a monomial of least degree appearing in g, then it cannot be cancelled out of the right-hand side of (17). On the other hand, it is easy to see that $\sum x_i$ is a non-zero-divisor in $\mathbb{R}[\Delta]$. Hence if the right-hand side of of (17) is written as a linear combination of monomials, then some monomial will appear of degree greater than the degree of v. This contradicts the homogeneity of the left-hand side, so the claim is proved.

If now g is a homogeneous element of $\mathbb{R}[\Delta]$ of degree $i < j$, then $h := (x_1 + \cdots + x_n)^{j-i}g$ is an element of $\mathbb{R}[\Delta]$ satisfying $\psi(g) = \psi(h)$. Hence

$$\psi(\mathbb{R}[\Delta]_j) = \psi(\mathbb{R}[\Delta]_0 \oplus \cdots \oplus \mathbb{R}[\Delta]_j),$$

and the proof follows. □

We can now deduce a formula for the dimension of the space $C_j^0(\Delta)$ in terms of the f-vector of Δ.

5.4 Corollary. *The dimension of the vector space $C_j^0(\Delta)$ is given by*

$$\dim C_j^0(\Delta) = \dim \mathbb{R}[\Delta]_j = \sum_{i=0}^{m} f_i(\Delta)\binom{j-1}{i},$$

for $j > 0$.

Proof. According to Theorem 5.2 and Lemma 5.3, the composite map $\varphi \circ \psi$ restricts to a bijection between $\mathbb{R}[\Delta]_j$ and $C_j^0(\Delta)$. The proof follows from Theorem II.1.4. □

Corollary 5.4 shows that $\dim C_j^0(\Delta)$ depends only on Δ as an *abstract* simplicial complex (in fact, only on the f-vector), and not on the embedding of Δ into $\mathbb{R}^m$. This phenomenon fails for $r > 0$ and greatly complicates the situation in this case.

A final topic considered by Billera in [18] is the structure of $C^0[\Delta]$ as a module over the polynomial subring R generated by the coordinate functions $y_1, \ldots, y_m$. Define $\theta_i \in \mathbb{R}[\Delta]_1$ for $1 \le i \le m$ by

$$\theta_i = \sum_{x \in V} y_i(x)x,$$

where the sum ranges over the vertex set $V \subset \mathbb{R}^m$ of Δ (so $y_i(x)$ is the ith coordinate of x). By Lemma 2.4(a), $\theta_1, \ldots, \theta_m$ is an l.s.o.p. for $\mathbb{R}[\Delta]$. Hence if $\mathbb{R}[\Delta]$ is Cohen–Macaulay, then $\mathbb{R}[\Delta]$ is a finitely-generated free $\mathbb{R}[\theta_1, \ldots, \theta_m]$-module; and if in addition $\mathbb{R}[\Delta]$ is shellable then we can write down a set of basis elements from the shelling (Theorem 2.5). The image $\psi\varphi(\theta_i)$ of θ_i in $C^0(\Delta)$ is just y_i. From this it is easy to carry over the results about the structure of $\mathbb{R}[\Delta]$ as a $\mathbb{R}[\theta_1, \ldots, \theta_m]$-module to the structure of $C^0(\Delta)$ as an R-module. We obtain that $C^0(\Delta)$ is a free R-module if Δ is Cohen–Macaulay (over $\mathbb{R}$). In fact, Billera and Rose have shown [24, Thm. 4.5] that we only need for Δ to be Buchsbaum, i.e., Δ is pure, and the link of every vertex is Cohen–Macaulay. For instance, any triangulation of a manifold (with or without boundary) is Buchsbaum. If moreover Δ is shellable, then the shelling yields an explicit basis for $C^0(\Delta)$ as an R-module, consisting of certain squarefree products of Courant functions. See [18, §4] for further details.

6 Algebras with straightening law and simplicial posets

The theory of face rings has applications to even broader classes of rings. The basic reason for this is that certain rings R can be "approximated" by a face ring $k[\Delta]$, and many of the algebraic properties of $k[\Delta]$ carry over to R. The most satisfactory class of rings which can be treated in this way are called "algebras with straightening law" (ASL's). The basic reference to this subject is [62]. Other references include [8], [11], [48, Ch. 7.1], [72], [80], [81], [82], [91, Part 3]. Our treatment will for simplicity not be done in the greatest possible generality as in [62]. Instead we will follow [91, Part 3]. In [62] the terms "algebra with straightening law" and "Hodge algebra" are used synonymously, but in [91] Hodge algebras are more general than ASL's. A closely related concept is that of Gröbner bases. We will not treat Gröbner bases here, but we should at least mention that they are the key tool in the subject of computational commutative algebra. See for instance [1], [14], [49], [58], [177]. We will give one main combinatorial application of ASL's to a generalization of a simplicial complex known as a "simplicial poset."

Let us proceed to the definition of an ASL (over a field k). Suppose P is a finite poset. A *monomial* $\mathcal{M}$ on P is a function $\mathcal{M} : P \to \mathbb{N}$. The *support* of $\mathcal{M}$ is the set

$$\mathrm{supp}(\mathcal{M}) = \{y \in P : \mathcal{M}(y) \neq 0\}.$$

A monomial $\mathcal{M}$ is *standard* if $\mathrm{supp}(\mathcal{M})$ is a chain of P. If R is a commutative ring with an injection $\phi : P \to R$ given, then to each monomial $\mathcal{M}$ on P we may associate

$$\phi(\mathcal{M}) := \prod_{y \in P} \phi(y)^{\mathcal{M}(y)} \in R.$$

We will identify P with $\phi(P)$ and will also call $\phi(\mathcal{M})$ a monomial.

Now let k be a field and R a (commutative) k-algebra in the situation of the previous paragraph. We will call R (or more accurately the pair

$(R, \phi))$ an *algebra with straightening law* (ASL) on P over k if the following two conditions are satisfied:

(ASL-1) The set of standard monomials is a basis of the algebra R as a vector space over k.

(ASL-2) If x and y in P are incomparable, and if

$$xy = \sum_\imath a_\imath x_{\imath 1} x_{\imath 2} \cdots x_{\imath r_\imath}, \qquad (18)$$

where $0 \neq a_i \in k$ and $x_{\imath 1} \leq x_{\imath 2} \leq \cdots$, is the unique expression for $xy \in R$ as a linear combination of standard monomials guaranteed by (ASL-1), then $x_{\imath 1} \leq x$ and $x_{i1} \leq y$ for every i.

Note that the right-hand side of (18) is allowed to be the empty sum $(= 0)$; but that, though 1 is a standard monomial (whose support is the empty chain), no product $x_{\imath 1} x_{\imath 2} \cdots x_{ir_\imath}$ can be 1.

Suppose that R is an ASL on P, and that in addition R is a connected graded algebra with $P \subset \mathcal{H}(R_+)$, i.e., every element of P (regarded as a subset of R) is homogeneous of positive degree. We then say that R is a *graded* ASL.

The main result about ASL's in which we're interested is the following [48, Cor. 7.1.6], [62, Cor. 7.2]. It is a basic tool for proving that certain rings are Cohen–Macaulay. The proof is rather technical and is not included here.

6.1 Theorem. *Let P be a finite Cohen–Macaulay poset and R a graded ASL on P over k. Then R is Cohen–Macaulay.*

The above theorem easily extends to the result that if R is an ASL on an arbitrary finite poset P over k, and if R is an $\mathbb{N}$-graded connected k-algebra such that $P \subset \mathcal{H}R_+$, then $\text{depth}(R) \geq \text{depth}(k[P])$.

6.2 Example. Let $[x_{ij}]$ be an $m \times n$ matrix of indeterminates over k with $m \leq n$, and set G_{mn} equal to the subalgebra of the polynomial ring $k[x_{ij}]$ generated by all $m \times m$ minors of the matrix $[x_{ij}]$. The algebra G_{mn} is the homogeneous coordinate ring of the Grassmann variety of m-planes in n-space; see [98, Ch. XIV], [109, p. 1076], [153, §4]. It was shown essentially by Hodge [97] that the ring G_{mn} is an ASL in the following natural way. (For a modern treatment, see [61].)

Let $L(m, n - m)$ be the set of all symbols $[i_1, \ldots, i_m]$ with $1 \leq i_1 < i_2 < \cdots < i_m \leq n$, with the partial ordering $[i_1, \ldots, i_m] \leq [j_1, \ldots, j_m]$ if $i_r \leq j_r$ for all r. Define an injection $L(m, n - m) \hookrightarrow G_{mn}$ by taking $[i_1, \ldots, i_m]$ to the $m \times m$ minor of $[x_{ij}]$ which is the determinant of the submatrix of $[x_{ij}]$ consisting of columns $i_1, \ldots, i_m$. We will also write $[i_1, \ldots, i_m]$ for this minor.

The minors $[i_1, \ldots, i_m]$ form a minimal set of generators for G_{mn} (as a k-algebra). The relations among these generators are the *Plücker relations* [78, §9.1], [79, (15.53)], [98, (3) on p. 310], [109, (QP) on p. 1076]. One sees easily (see e.g. [98, Thm. I, p. 378]) that these relations imply the axioms (ASL-1) and (ASL-2) of an ASL, so that G_{mn} is an ASL on the poset $L(m, n - m)^3$.

The poset $L(m, n - m)$ is in fact a distributive lattice. Distributive lattices may be seen in several ways to be Cohen–Macaulay (see e.g. [34]), so from Theorem 6.1 we conclude that G_{mn} is Cohen–Macaulay. In fact, the poset of join-irreducible elements of $L(m, n-m)$ has the property that every maximal chain has the same length, from which it is immediate from Theorem i.12.7 and [146, Prop. 19.3] that G_{mn} is in fact *Gorenstein*.

The above technique for showing that G_{mn} is Cohen–Macaulay extends easily to the more general *Schubert varieties* (in the Grassmannian), whose homogeneous coordinate rings are ASL's on intervals $[\hat{0}, t]$ of $L(m, n - m)$ (where $\hat{0}$ denotes the unique minimal element $[1, 2, \ldots, m]$ of $L(m, n - m)$). Even more general are the *skew Schubert varieties* (the intersection of two Schubert varieties defined with respect to "opposite" flags), whose homogeneous coordinate rings are ASL's on arbitrary intervals $[s, t]$ of $L(m, n - m)$. For further information, see [155], [61], [62], [153], [178]. Additional classes of rings that can be proved to be Cohen–Macaulay by ASL techniques are discussed in [62] and the references cited there.

We now give a combinatorial application of Theorem 6.1. A *simplicial poset* [162, p. 135] (also called a *boolean poset* [82, p. 130] and a *poset of boolean type* [29, §2.3]) is a finite poset P with $\hat{0}$ such that every interval $[\hat{0}, y]$ is a boolean algebra (i.e., since P is finite, $[\hat{0}, y]$ is isomorphic to the set of all subsets of a finite set, ordered by inclusion). One may think of a simplicial poset as a generalization of a simplicial complex, such that if F and G are two faces then the intersection $2^F \cap 2^G$ of the corresponding simplices can be any subcomplex of the boundaries of 2^F and 2^G, and not just the single simplex $2^{F \cap G}$. For instance, Figure 3.2 shows a simplicial poset corresponding to two triangles which intersect in an edge together with the opposite vertex. Note that a simplicial poset P is the face poset of a simplicial complex if and only if P is a meet-semilattice, i.e., any two

---

[3]For some other connections between the poset $L(m, n-m)$ and the Grassmannian, see for instance [153, p. 239] and [157].

Figure 3.2: A simplicial poset

elements x and y have a greatest lower bound or *meet* $x \wedge y$. In general, a simplicial poset is always the face poset of a regular CW complex (see [29]).

We can define the f-vector and h-vector of a simplicial poset P in exact analogy to simplicial complexes. Namely, define the f-vector by

$$f_i = f_i(P) = \#\{y \in P : [\hat{0}, y] \cong B_{i+1}\}, \tag{19}$$

where B_{i+1} denotes the boolean algebra of rank $i + 1$. If P has rank d (i.e., $d = \max\{i : \exists y \in P \text{ such that } [\hat{0}, y] \cong B_i\}$), we then define $h_i = h_i(P)$ just as in equation (4). In particular, if P is the face poset of a simplicial complex Δ then $(f_0, f_1, \ldots, f_{d-1})$ is just the f-vector of Δ, and $(h_0, h_1, \ldots, h_d)$ is just the h-vector.

Our reason for dealing with simplicial posets P is that one can define a graded algebra A_P which is a direct generalization of the face ring of a simplicial complex and which shares many properties with the face ring. When P is the face poset of a simplicial complex Δ then indeed $A_P \cong k[\Delta]$.

6.3 Definition. Let P be a simplicial poset with elements $\hat{0} = y_0, y_1, \ldots, y_p$. Let $A = k[y_0, y_1, \ldots, y_p]$ be the polynomial ring over k in the variables y_i. Define I_P to be the ideal of A generated by the following elements:

(S1) $y_i y_j$, if y_i and y_j have no common upper bound in P,

(S2) $y_i y_j - (y_i \wedge y_j) (\sum_z z)$, where z ranges over all *minimal* upper bounds of y_i and y_j, otherwise.

Finally set $\tilde{A}_P = A/I_P$ and $A_P = \tilde{A}_P/(y_0 - 1)$. We call A_P the *face ring* of the simplicial poset P. We make A_P into a graded algebra $A_P = (A_P)_0 \oplus (A_P)_1 \oplus \cdots$ by defining $\deg(y_i) = j$ if $[\hat{0}, y] \cong B_j$, for $0 \le i \le p$.

NOTE. (a) It is clear that $y_i \wedge y_j$ exists whenever y_i and y_j have at least one upper bound z since the interval $[\hat{0}, z]$ is a boolean algebra (and therefore a lattice). Hence property (S2) is well-defined.

(b) We could replace $\sum_z z$ in the above definition by $\sum_z \alpha_z z$ for any $0 \neq \alpha_z \in k$ without any significant change in the theory.

(c) The grading we have given to A_P is well-defined since the relations (S2) and $y_0 - 1$ are easily seen to be homogeneous. Moreover, $\dim_k(A_P)_0 = 1$, so A_P is connected.

Our main goal here is to characterize the f-vector (or h-vector) of a Cohen–Macaulay simplicial poset. (For the much easier problem of characterizing the f-vector of an arbitrary simplicial poset, see [165, Thm. 2.1].) We will just sketch the main ideas; complete details appear in [165, §3].

- *Step 1.* An elementary argument shows that $\tilde{A}_P$ is an ASL on P.

- *Step 2.* It follows from Theorem 6.1 and Step 1 above that if P is a finite Cohen–Macaulay simplicial poset, then the ring $\tilde{A}_P$ is Cohen–Macaulay.

- *Step 3.* Let P be a finite poset with $\hat{0} = y_0$. Let A_P be an ASL on P. Then $y_0 - 1$ is a non-zero-divisor of A_P. (For if u is a standard monomial then so is $y_0 u$. Hence if $0 \neq z \in A_P$ and $(y_0 - 1)z = 0$, then we get a nontrivial relation among standard monomials, contradicting (ASL-1).)

- *Step 4.* Let $P = \{y_0, y_1, \ldots, y_p\}$ be a finite Cohen–Macaulay simplicial poset with $y_0 = \hat{0}$. Combining Step 2, Step 3, and the standard result [107, Theorem 141] that if a ring R is Cohen–Macaulay and $y \in R$ is neither a zero-divisor nor a unit then R/yR is Cohen–Macaulay, we get that the ring $A_P = \tilde{A}_P/(y_0 - 1)$ is Cohen–Macaulay.

- *Step 5.* Let P be a finite simplicial poset. A simple counting argument shows that the Hilbert series of A_P is given by

$$F(A_P, \lambda) = \sum_{x \in P} \frac{\lambda^{\rho(x)}}{(1 - \lambda)^{\rho(x)}} = \sum_{i=0}^{d} f_{i-1} \frac{\lambda^i}{(1 - \lambda)^i},$$

where $\rho(x)$ denotes the rank of x. It follows immediately from (4) that

$$F(A_P, \lambda) = \frac{h_0 + h_1\lambda + \cdots + h_d\lambda^d}{(1 - \lambda)^d}. \tag{20}$$

- *Step 6.* We want to show that A_P has an l.s.o.p. (when k is infinite). Since A_P is not standard (i.e., not generated by elements of degree one), we cannot simply appeal to Lemma i.5.2. Instead one shows by a simple induction argument (given in [165, Lemma 3.9]) that if A_P^* is the subalgebra of A_P generated by elements of degree one, then A_P is integral over A_P^*. Hence A_P is a finitely-generated A_P^*-module, so every h.s.o.p for A_P^* is an h.s.o.p. for A_P. It now follows from Lemma i.5.2 that A_P has an l.s.o.p.

- *Step 7.* Let P be simplicial and Cohen–Macaulay of rank d. If $\theta_1, \ldots, \theta_d$ is an l.s.o.p. for A_P (existence guaranteed by Step 6), then by (20) and the discussion in Chapter i.5, we have that

$$F(A_P/(\theta_1, \ldots, \theta_d), \lambda) = h_0 + h_1\lambda + \cdots + h_d\lambda^d.$$

Hence $h_i \geq 0$.

- *Step 8.* Conversely, given a sequence $(h_0, \ldots, h_d)$ of nonnegative integers with $h_0 = 1$, one can construct by a shelling argument [165, p. 326] a Cohen–Macaulay simplicial poset P of rank d and h-vector $(h_0, \ldots, h_d)$.

Putting the above discussion together, we obtain a characterization of h-vectors (or equivalently, of f-vectors) of Cohen–Macaulay simplicial posets [165, Thm. 3.10].

6.4 Theorem. *Let* $\mathbf{h} = (h_0, h_1, \ldots, h_d) \in \mathbb{Z}^{d+1}$. *The following two conditions are equivalent:*

(i) *There exists a Cohen–Macaulay simplicial poset P of rank d with h-vector $h(P) = \mathbf{h}$.*

(ii) $h_0 = 1$, *and* $h_i \geq 0$ *for all* i.

An interesting open problem is the characterization of h-vectors of *Gorenstein* simplicial posets. The only Gorenstein simplicial posets P for which $P - \{\hat{0}\}$ is acyclic are the boolean algebras, so we may as well assume $P - \{\hat{0}\}$ is nonacyclic. Equivalently, $P \cup \{\hat{1}\}$ is Cohen–Macaulay and Eulerian. The Dehn-Sommerville equations (Chapter II.6) generalize easily to the present setting (see [165, Prop. 4.4] for a more general result) as follows.

6.5 Proposition. *Let P be a Cohen–Macaulay simplicial poset of rank d for which $P \cup \{\hat{1}\}$ is Eulerian. If the h-vector of P (as defined using (19) and not to be confused with the h-vector of the order complex $\mathcal{O}(P)$) is $(h_0, h_1, \ldots, h_d)$, then $h_i = h_{d-i}$ for $0 \leq i \leq d$.*

Thus necessary conditions on the h-vector of a Gorenstein simplicial poset of rank d that is not a boolean algebra are that $h_0 = 1$, $h_i \geq 0$, and $h_i = h_{d-i}$. These conditions are not far from being sufficient. In fact, it is shown in [165, Thm. 4.3] that these conditions are sufficient if P has an even number of elements of rank d (equivalently, $h_0 + h_1 + \cdots + h_d$ is even, which is always the case when d is odd by Proposition 6.5) or if $h_i > 0$ for $0 \leq i \leq d$. Unfortunately the above necessary conditions are not sufficient in general. For instance, it follows from [165, Thm. 4.5] that $(1, 0, 1, 0, 1)$ is not a valid h-vector. For some additional necessary conditions, see also [67] and [165, Prop. 4.6].

7   Relative simplicial complexes

A *relative simplicial complex* is a collection Ψ (always assumed to be finite) of finite sets such that if $F, G \in \Psi$ and $F \subset H \subset G$, then $H \in \Psi$. Equivalently, there is a simplicial complex Δ and a subcomplex Γ for which $\Psi = \Delta - \Gamma$. We will often write this as $\Psi = \Delta/\Gamma$. Much of the connection between commutative algebra and simplicial complexes can be "relativized." We will only mention a few highlights here. The first step is to define the analogue of the face ring. Namely, given the relative complex $\Psi = \Delta/\Gamma$, define $I_\Psi = I_{\Delta/\Gamma}$ to be the ideal of $k[\Delta]$ generated by all face monomials x^F for which $F \notin \Gamma$. We may regard $I_{\Delta/\Gamma}$ as a module over $k[\Delta]$ or over the polynomial ring $k[V]$ in the vertices of Δ.

The dimension, f-vector, and h-vector of a relative complex Ψ are defined exactly analogously to the ordinary case of simplicial complexes. Thus,

$$\dim(\Psi) = d - 1 = \max\{|F| - 1 : F \in \Psi\}$$
$$f_i(\Psi) = \#\{F \in \Psi : |F| = i + 1\}$$
$$\sum_{i=0}^{d} f_{i-1}(\Psi)(x-1)^{d-i} = \sum_{i=0}^{d} h_i(\Psi)x^{d-i}.$$

The face module $I_{\Delta/\Gamma}$ inherits a grading from $k[\Delta]$. The results in Chapter II.1 and II.2 on the Hilbert series of the face ring of a simplicial complex carry over immediately to the relative case. Specifically, we have the following result.

7.1 Proposition. *Let Ψ be a $(d-1)$-dimensional relative complex with f-vector $(f_0, \ldots, f_{d-1})$ and h-vector $(h_0, \ldots, h_d)$. Then*

$$F(I_\Psi, \lambda) = \sum_{i=0}^{d} f_{i-1} \frac{\lambda^i}{(1-\lambda)^i}$$

$$= \frac{\sum_{i=0}^{d} h_i \lambda^i}{(1-\lambda)^d}. \tag{21}$$

If I_Ψ is a Cohen–Macaulay module (in which case we call Ψ a *Cohen–Macaulay relative simplicial complex*), then it follows from Chapter i.5 and equation (21) that $h_i(\Psi) \geq 0$. There is an exact analogue of Reisner's theorem (Corollary II.4.2) for the relative case, proved in exactly the same way we proved Corollary II.4.2. To state this result (which first appeared in [164, Thm. 5.3]), we let $\tilde{H}_i(\Delta, \Gamma; k)$ denote the ith reduced relative homology group of the pair (Δ, Γ) over the field k. It is important to distinguish between the cases $\Gamma = \emptyset$ (the empty simplicial complex) and $\Gamma = \{\emptyset\}$ (the simplicial complex with the unique face $\emptyset$, of dimension -1). We then have

$$\tilde{H}_i(\Delta, \emptyset; k) \cong \tilde{H}_i(\Delta; k)$$

$$\tilde{H}_i(\Delta, \{\emptyset\}; k) \cong H_i(\Delta; k).$$

Furthermore, for $F \in \Delta$ define $\mathrm{lk}_\Gamma F = \{G \in \Gamma : F \cap G = \emptyset, F \cup G \in \Gamma\}$.

7.2 Theorem. *The ideal $I_{\Delta/\Gamma}$ is Cohen–Macaulay if and only if for all $F \in \Delta$ (including $F = \emptyset$), we have*

$$\tilde{H}_i(\mathrm{lk}_\Delta F, \mathrm{lk}_\Gamma F; k) = 0, \quad \text{if } i < \dim(\mathrm{lk}_\Delta F). \quad \Box$$

Some elementary consequences of Theorem 7.2 are collected in the next corollary.

7.3 Corollary. (i) *The property of being relative Cohen–Macaulay is topological, i.e., depends only on the geometric realization $|\Delta/\Gamma|$ (and the underlying field k).*

(ii) *Let Δ/Γ be Cohen–Macaulay, and suppose that F is a maximal face of Γ. Then $\dim(\mathrm{lk}_\Delta F) = -1$ or 0 (i.e., F is a maximal face of Δ or a codimension one face of a maximal face of Δ).*

(iii) *If Δ triangulates a $(d-1)$-ball and Γ triangulates a $(d-2)$-ball contained in $\partial\Delta$, then Δ/Γ is Cohen–Macaulay.*

(iv) *If Δ is Cohen–Macaulay and Γ is Cohen–Macaulay of the same dimension, then Δ/Γ is Cohen–Macaulay.*

Proof. For (i)–(iii), see [164, Cor. 5.4]. (iv) is an immediate consequence of the long exact sequence for relative homology [130, Thm. 23.3]. $\quad \Box$

The theory of shellability developed in Section 2 generalizes straightforwardly to the relative case.

7.4 Definition. A pure relative simplicial complex Ψ is *shellable* if its facets can be ordered $F_1, F_2, \ldots, F_s$ such that if Δ_j denotes the relative subcomplex

$$\Delta_j = \left(2^{F_1} \cup 2^{F_2} \cup \cdots \cup 2^{F_j}\right) \cap \Psi,$$

then for all $1 \leq i \leq s$, the set of faces of Δ_i which do not belong to Δ_{i-1} has a unique minimal element (with respect to inclusion). The linear order $F_1, \ldots, F_s$ is then called a *shelling order* or a *shelling* of Ψ.

Given a shelling $F_1, \ldots, F_s$ of Ψ, we define the *restriction* $r(F_i)$ exactly as in Section 2, and similarly we define a *partitioning* of Ψ as in Section 2. Proposition 2.3 and Theorem 2.5 carry over, *mutatis mutandis*, for relative complexes. Hence shellable relative complexes are Cohen–Macaulay. Moreover, Conjecture 2.7 may be extended to the relative case.

For some additional results related to the combinatorics of relative complexes, see [68]. We conclude this section with an example of a Cohen–Macaulay relative simplicial complex related to the combinatorics of finite posets. The basic combinatorial theory on which this example is based may be found in [146]. Let P be a p-element poset and $\omega : P \to [p] = \{1, 2, \ldots, p\}$ a bijection. The pair (P, ω) is called a *labelled poset*. Let $J(P)$ denote the lattice of order ideals of P [162, p. 100 and Ch. 3.4], i.e.,

$$J(P) = \{I \subseteq P : t \in I, s < t \Rightarrow s \in I\}.$$

Order the elements of $J(P)$ by inclusion. Let $\tilde{J}(P) = J(P) - \{\emptyset, P\}$, i.e., $\tilde{J}(P)$ consists of $J(P)$ with the top and bottom elements removed. Let $\Psi = \Psi(P, \omega)$ be the set of all chains $I_1 < I_2 < \cdots < I_t$ in $\tilde{J}(P)$ such that in each of the subposets $I_1, I_2 - I_1, \ldots, I_t - I_{t-1}, P - I_t$ of P, the labelling ω is *order-preserving*. Clearly Ψ is a relative simplicial complex (contained in the order complex $\mathcal{O}(\tilde{J}(P))$). We will simply state here a few of the basic properties of $\Psi(P, \omega)$.

- $\Psi(P, \omega)$ is Cohen–Macaulay. Two proofs are given in [172, Prop. 4.2]. The first shows that the geometric realization of Ψ is either a sphere of dimension $p - 2$ (when P is totally disconnected), or else a ball of dimension $p - 2$, from the boundary of which is removed either nothing (when ω is order-preserving), the entire boundary (when ω is order-reversing), or a ball of dimension $p - 3$ (in all other cases). Now use Corollary 7.3(ii). The second proof is based on a relative version of the theory of lexicographic shellability.

- Let $\pi : P \to [p]$ be a *linear extension* (order-preserving bijection), and let $a_i = \omega(\pi^{-1}(i))$. Let w_π denote the permutation $a_1, a_2, \ldots, a_p$ of $[p]$. Define

$$d(w_\pi) = \#\{i : a_i > a_{i+1}\},$$

  the number of *descents* of w_π. Then the h-vector $(h_0, h_1, \ldots, h_{p-1})$ of $\Psi(P, \omega)$ is given by

$$\sum_{i=0}^{p-1} h_i x^i = \sum_\pi x^{d(w_\pi)},$$

  the latter sum being over all linear extensions π of P.

- Let $H(\Psi, n)$ denote the Hilbert function of the ideal I_Ψ of the face ring $k[\mathcal{O}(\bar{J}(P))]$. Then for $n \geq 1$, $H(\Psi, n-1)$ is the number of order-preserving maps $\sigma : P \to [n]$ such that if $s < t$ in P and $\omega(s) > \omega(t)$, then $\sigma(s) < \sigma(t)$.

- Let $\bar{\omega} : P \to [p]$ be defined by $\bar{\omega}(t) = p + 1 - \omega(t)$ for all $t \in P$. Then the canonical module $\Omega(I_{\Psi(P,\omega)})$ of the ideal $I_{\Psi(P,\omega)}$, as defined in Chapter i.12, is given by

$$\Omega(I_{\Psi(P,\omega)}) \cong I_{\Psi(P,\bar{\omega})}.$$

The proof is a consequence of [146, Thm. 10.1] and [160, Thm. 4.4].

8 Group actions

In this section we discuss how under certain circumstances the action of a group G on a simplicial complex Δ leads to a refinement of the $\mathbb{N}$-grading of the face ring, which in turn allows us to obtain information on the f-vector of Δ. The most interesting case is when G has order two, so for simplicity we will only treat this case here.

The original motivation for this work arose from the paper [13] of Bárány and Lovász. They showed that a centrally-symmetric simplicial d-polytope $\mathcal{P}$ has at least 2^d facets. ("Centrally symmetric" means that $z \in \mathcal{P}$ if and only if $-z \in \mathcal{P}$.) Equality is achieved by the d-dimensional cross-polytope (the dual to the d-cube). They also conjectured for centrally-symmetric simplicial d-polytopes with $2n$ vertices explicit lower bounds for each f_i. In terms of the h-vector, the inequality $f_{d-1} \geq 2^d$ of Bárány-Lovász takes the form

$$h_0 + h_1 + \cdots + h_d \geq 2^d.$$

This inequality led Björner (unpublished) to conjecture that in fact $h_i \geq \binom{d}{i}$ for all i. Even more strongly (since $h_0 = 1$), Björner conjectured that

$$h_i - h_{i-1} \geq \binom{d}{i} - \binom{d}{i-1}, \quad 1 \leq i \leq \lfloor d/2 \rfloor, \tag{22}$$

a strengthening of the inequality $h_i - h_{i-1} \geq 0$ for $1 \leq i \leq \lfloor d/2 \rfloor$ satisfied by any simplicial d-polytope by Theorem 1.1. It is not difficult to see that the inequalities (22) imply the conjecture of Bárány and Lovász. In this section we will indicate how techniques from commutative algebra imply Björner's conjecture $h_i \geq \binom{d}{i}$ for simplicial complexes considerably more general than boundary complexes of centrally-symmetric simplicial d-polytopes. We will also briefly indicate how the techniques of Section 1 can then be employed to prove (22) for the originally conjectured case of boundary complexes of centrally symmetric simplicial d-polytopes.

Suppose that G is a group of order two, say with elements 1 and σ. Henceforth in this section we will assume that the ground field k satisfies char $k \neq 2$. If G acts on a vector space W over k then we have a direct sum decomposition $W = W^+ \oplus W^-$, where

$$W^+ = \{v \in W : \sigma(v) = v\}, \quad W^- = \{v \in W : \sigma(v) = -v\}.$$

In terms of representation theory, this is just the decomposition of W into its isotypic components under the action of G. However, G is such a simple group that we can deal with it without explicitly invoking representation theory.

Now suppose that G acts on a simplicial complex Δ, i.e., we have a homomorphism $G \to \mathrm{Aut}(\Delta)$, the automorphism group of Δ. The action of G on Δ extends in an obvious way to an action on $k[\Delta]$. The decomposition $k[\Delta] = k[\Delta]^+ \oplus k[\Delta]^-$ defines a $\mathbb{Z}/2\mathbb{Z}$-grading on $K[\Delta]$ (i.e., $k[\Delta]^+ \cdot k[\Delta]^+ \subseteq k[\Delta]^+$, etc.). Since each homogeneous component $k[\Delta]_i$ is G-invariant, it follows that $k[\Delta]$ has in fact a natural $\mathbb{N} \times (\mathbb{Z}/2\mathbb{Z})$-grading.

The main idea of this section is to exploit the $\mathbb{N} \times (\mathbb{Z}/2\mathbb{Z})$-grading when Δ is Cohen–Macaulay by looking at the quotient of $k[\Delta]$ by a homogenous (with respect to the $\mathbb{N} \times (\mathbb{Z}/2\mathbb{Z})$-grading) linear system of parameters. In general such an l.s.o.p. will not exist, so we must impose additional conditions on the group action. If G is any finite group of automorphisms of Δ, then we say that G acts *very freely* on Δ if for every $\sigma \neq 1$ in G and every vertex x of Δ, we have that $x \neq \sigma(x)$ and that $\{x, \sigma(x)\}$ is not an edge of Δ. Equivalently, for every vertex x, the open stars of the elements of the orbit Gx are pairwise disjoint. We say that G acts *fixed-point freely* or just *freely* on Δ if for every $\sigma \neq 1$ in G and every face $F \neq \emptyset$ of Δ, we have that $\sigma(F) \neq F$. Equivalently, in the affine extension of this action to the geometric realization $|\Delta|$ of Δ, only the identity element of G has any fixed points. Note that a very free action is always free, but not conversely (e.g., the group of order three acting in the obvious way on the boundary of a triangle). When G has order two, free actions and very free actions are the same.

The most obvious example of a $(d-1)$-dimensional Cohen–Macaulay simplicial complex Δ with a free involution (i.e., the free action of a group of order two) is the boundary complex of the d-dimensional cross-polytope. If we take the vertices of Δ to be the integers $\pm 1, \pm 2, \ldots, \pm d$ with i and $-i$ antipodal, then the facets consist of all d-element subsets of vertices not containing both i and $-i$. The h-vector of Δ is easily computed to be $(1, d, \binom{d}{2}, \binom{d}{3}, \ldots, 1)$. The next theorem shows that this is the smallest possible h-vector among all $(d-1)$-dimensional Cohen–Macaulay complexes with a free involution. As mentioned above, this inequality was conjectured by Björner (unpublished) for the special case of centrally-symmetric simplicial polytopes (in which case the antipodal map induces a free involution on the boundary complex).

8.1 Theorem. *Let Δ be a $(d-1)$-dimensional Cohen–Macaulay simplicial complex, and suppose that a group G of order two acts freely on Δ. Then*

$$h_i(\Delta) \geq \binom{d}{i}, \quad 0 \leq i \leq d.$$

In particular,

$$f_{d-1} = h_0 + h_1 + \cdots + h_d \geq 2^d.$$

Proof (sketch). The idea is to choose a certain l.s.o.p. of $k[\Delta]$ which generates a G-invariant ideal I, and to compute the Hilbert function of $k[\Delta]/I$ with respect to the $\mathbb{N} \times (\mathbb{Z}/2\mathbb{Z})$-grading. We want the l.s.o.p. $\theta_1, \ldots, \theta_d$ to have the property that $\theta_i \in k[\Delta]_1^-$ (for reasons soon to become clear). To construct such an l.s.o.p., choose a subset V' of the vertex set V of

Δ such that V' contains one element from each G-orbit of V. Since we are assuming k is infinite, there exist functions $t_1, \ldots, t_d : V' \to k$ such that the restrictions of $t_1, \ldots, t_d$ to any d-element subset of V' are linearly independent. Extend $t_1, \ldots, t_d$ to all of V by defining $t_i(\sigma(x)) = -t_i(x)$ for $x \in V'$. Define

$$\theta_i = \sum_{x \in V} t_i(x)x.$$

Clearly $\theta_i \in k[\Delta]_1^-$. Since x and $\sigma(x)$ are not both vertices of any face of Δ, it follows from Lemma 2.4(a) that $\theta_1, \ldots, \theta_d$ form an l.s.o.p., as desired.

Let $A = k[\Delta]/(\theta_1, \ldots, \theta_d)$, with $\theta_1, \ldots, \theta_d$ as above. Since $\sigma(\theta_i) = -\theta_i$, it follows that the ideal $(\theta_1, \ldots, \theta_d)$ is G-invariant, and hence that G acts on the graded algebra $A = A_0 \oplus \cdots \oplus A_d$. In other words, A has an $\mathbb{N} \times (\mathbb{Z}/2\mathbb{Z})$-grading. If W is any $\mathbb{N} \times (\mathbb{Z}/2\mathbb{Z})$-graded k-vector space, then define the Hilbert series $F(W, \lambda, t)$ of W by

$$F(W, \lambda, t) = \sum_{i \geq 0} \left[(\dim_k W_i^+) + (\dim_k W_i^-)t \right] \lambda^i,$$

where λ and t are indeterminates with $t^2 = 1$. It is clear that for $i \geq 1$,

$$\dim_k k[\Delta]_i^+ = \dim_k k[\Delta]_i^- = \frac{1}{2} \dim_k k[\Delta]_i. \qquad (23)$$

For $i = 0$, on the other hand, we have

$$\dim_k k[\Delta]_0^+ = 1, \quad \dim_k k[\Delta]_0^- = 0. \qquad (24)$$

It follows from the definition of the h-vector $h(\Delta)$ in Chapter II.5 that the usual $\mathbb{N}$-graded Hilbert series $F(k[\Delta], \lambda)$ of $k[\Delta]$ is given by

$$F(k[\Delta], \lambda) = \frac{h_0 + h_1\lambda + \cdots + h_d\lambda^d}{(1 - \lambda)^d}. \qquad (25)$$

It is an immediate consequence of equations (23), (24), and (25) that

$$F(k[\Delta], \lambda, t) = 1 + \frac{1}{2} \left[\frac{\sum_{i=0}^d h_i\lambda^i}{(1 - \lambda)^d} - 1 \right] (1 + t). \qquad (26)$$

Each time we divide out by a parameter θ_i we multiply the Hilbert series by $1 - \lambda t$, so

$$\begin{aligned} F(A, \lambda, t) &= (1 - \lambda t)^d F(k[\Delta], \lambda, t) \\ &= \frac{(1 - \lambda t)^d}{2} \left[1 - t + (1 + t)(1 - \lambda)^{-d} \sum_{i=0}^d h_i\lambda^i \right]. \qquad (27) \end{aligned}$$

It is not difficult to simplify (27) to obtain

$$F(A, \lambda, t) = \frac{1}{2} \left[(1 - t)(1 + \lambda)^d + (1 + t) \sum_{i=0}^{d} h_i \lambda^i \right].$$

It follows that

$$\dim_k A_i^+ = \frac{1}{2} \left(h_i + \binom{d}{i} \right)$$

$$\dim_k A_i^- = \frac{1}{2} \left(h_i - \binom{d}{i} \right). \tag{28}$$

Since $\dim_k A_i^- \geq 0$, the proof follows. $\square$

It is natural to ask what other information can be obtained about the h-vector of a Cohen–Macaulay complex Δ with a free action of a group of order two. Ideally we would like a complete characterization analogous to Theorem II.3.3. At present such a definitive result is lacking. We will give one rather weak additional result which can be proved by commutative algebra techniques, and then will give an intriguing related conjecture. Let us also mention the further result, valid even if Δ is not Cohen–Macaulay and easily proved without algebraic machinery (see [163, Prop. 3.3]), that

$$h_i \equiv \binom{d}{i} \pmod 2. \tag{29}$$

8.2 Proposition. *Let Δ be a finite $(d-1)$-dimensional Cohen–Macaulay complex admitting a free $\mathbb{Z}/2\mathbb{Z}$-action. Suppose $h_i = \binom{d}{i}$ for some $i \geq 1$. Let $j \geq i$. If either j is even or $j - i$ is even, then $h_j = \binom{d}{j}$.*

Proof. The subspace A^+ of A is a subalgebra, and A^- is an A^+-module. Since $h_i = \binom{d}{i}$ we have $A_i^- = 0$ by (28). Suppose $j \geq i$ and $A_j^- \neq 0$. It is easy to check that A^+ is generated by A_1^+ and A_2^+ as a k-algebra, and that A^- is generated by A_1^- as an A^+-module. Hence there exist elements $t \in A_1^-$ and $u_1, \dots, u_r \in A_1^+ \cup A_2^+$ such that

$$0 \neq u_1 \cdots u_r t \in A_j^-.$$

If j is even then some $u_s \in A_1^+$. Then some subproduct v of $u_1 \cdots u_r$ will have degree $i - 1$. Hence $0 \neq vt \in A_i^-$, contradicting $\dim_k A_i^- = 0$. Similarly if $j - i$ is even then again some subproduct of $u_1 \cdots u_r$ will have degree $i - 1$, and we reach the same contradiction. $\square$

Proposition 8.2 suggests the following conjecture.

8.3 Conjecture. Let Δ be a finite $(d-1)$-dimensional Cohen–Macaulay complex admitting a free $(\mathbb{Z}/2\mathbb{Z})$-action. Suppose that $h_i = \binom{d}{i}$ for some $i \geq 1$. Then $h_j = \binom{d}{j}$ for all $j \geq i$.

We now indicate how Björner's conjecture (22) can be proved using the machinery of Section 1. The next result first appeared in [163, Thm. 4.1].

8.4 Theorem. *If $\mathcal{P}$ is a centrally-symmetric simplicial d-polytope with h-vector $(h_0, h_1, \ldots, h_d)$, then*

$$h_i - h_{i-1} \geq \binom{d}{i} - \binom{d}{i-1}, \quad 1 \leq i \leq \lfloor d/2 \rfloor.$$

Equivalently (using also the fact that $h_0 = 1$, the Dehn-Sommerville equations and equation (29)), the polynomial

$$\sum h_i x^i - (1+x)^d$$

has nonnegative symmetric and unimodal coefficients, all divisible by two.

Proof (sketch). Following the notation of Section 1, the description of the l.s.o.p. $\theta_1, \ldots, \theta_d$ of Theorem 1.3 (which is not given here but may be found e.g. in [60, §12], [77, §5.2]) is easily seen to have the form of the l.s.o.p. constructed in the proof of Theorem 8.1. The canonical action of the group $\mathbb{Z}/2\mathbb{Z}$ on $\mathcal{P}$ induces an action on the toric variety $X_\mathcal{P}$ and hence on the cohomology ring $H^*(X_\mathcal{P}, \mathbb{R})$, which coincides with the action on the graded algebra $A = k[\Delta]/(\theta_1, \ldots, \theta_d)$ used in the proof of Theorem 8.1. Moreover, this action commutes with multiplication by the hyperplane section ω. It follows that ω maps A_i^+ to A_{i+1}^+ and A_i^- to A_{i+1}^-. Since by Theorem 1.4 $\omega : A_{i-1} \to A_i$ is injective for $1 \leq i \leq \lfloor d/2 \rfloor$, we have

$$\dim_k A_{i-1}^- \leq \dim_k A_i^-, \quad 1 \leq i \leq \lfloor d/2 \rfloor.$$

The proof follows from (28). $\qquad\square$

It is natural to ask whether the theory presented thus far in this section can be extended to symmetry groups other than $\mathbb{Z}/2\mathbb{Z}$. The next result is a straightforward generalization of Theorem 8.1 and appeared first in [163, Prop. 3.3].

8.5 Proposition. *Let Δ be a $(d-1)$-dimensional Cohen–Macaulay complex with h-vector $(h_0, h_1, \ldots, h_d)$, and suppose that an abelian group G of order $g > 1$ acts very freely on Δ. Then*

$$h_i \geq \binom{d}{i}, \qquad i \text{ even}$$

$$h_i \geq (g-1)\binom{d}{i}, \quad i \text{ odd.}$$

In particular,

$$f_{d-1} = h_0 + h_1 + \cdots + h_d \geq g \cdot 2^{d-1}.$$

Proposition 8.5 suffers from the defect that it is not tight (best possible) when $g \geq 3$. Some interesting work of R. Adin [2], [3], [4] shows that better results are possible when we consider free actions rather than very free actions. We will simply state his main results here. The proofs use techniques similar to those used to prove Theorems 8.1 and 8.4, though the proof of the existence of a suitable l.s.o.p. is considerably more difficult.

8.6 Theorem. [2, Thm. 3.3] *Let Δ be a $(d-1)$-dimensional Cohen–Macaulay complex with a free action of the group $\mathbb{Z}/p\mathbb{Z}$, where p is prime and $p-1$ divides d. Let $h(\Delta) = (h_0, h_1, \ldots, h_d)$. Then*

$$\sum_{i=0}^{d} h_i x^i \geq (1 + x + \cdots + x^{p-1})^{d/(p-1)}, \qquad (30)$$

where the inequality holds coefficientwise. Moreover, the difference between the two sides of (30) has all coefficients divisible by p. The inequality (30) is best possible for all p and d satisfying the hypotheses.

Adin [2, Ch. 4] also has results when $p-1$ does not divide d, but they are not in general best possible. For the next result, let $\mathcal{P}$ be a convex d-polytope in $\mathbb{R}^d$, and let p be a prime. Let ν be a positive integer and $n = p^\nu$. A p^ν-*symmetry* of $\mathcal{P}$ is a linear transformation $T \in \mathrm{GL}(d, \mathbb{R})$ such that T maps $\mathcal{P}$ onto itself, $T^n = I$ (the identity transformation), and the origin is the only point of $\mathbb{R}^d$ fixed by $T^{n/p}$ (or equivalently, no eigenvalue of $T^{n/p}$ is equal to one). We do not assume that T is an orthogonal transformation with respect to the standard scalar product on $\mathbb{R}^d$. We say that $\mathcal{P}$ is *rational* if every vertex has rational coordinates.

8.7 Theorem. [3, Thm. 1] *Let $\mathcal{P}$ be a rational simplicial d-polytope with a p^ν-symmetry, for some prime p and integer $\nu \geq 1$. Let $h(\mathcal{P}) = (h_0, h_1, \ldots, h_d)$. Then*

(a) $d \equiv 0 \pmod{p^{\nu-1}(p-1)}$

(b) *The polynomial*

$$\sum_{i=0}^{d} h_i x^i - (1 + x + x^2 + \cdots + x^{p-1})^{d/(p-1)}$$

has symmetric and unimodal nonnegative coefficients, all of which are divisible by p^ν.

For some further related work and open problems, see the three papers of Adin. It seems quite likely that Theorem 8.6 has a extension to prime powers analogous to Theorem 8.7, but at present this remains open.

9 Subcomplexes

What can be said about the f-vector of a subcomplex Γ of a simplicial complex Δ? Here we will be concerned with the h-vector of a Cohen–Macaulay subcomplex of a Cohen–Macaulay complex. (For the problem of characterizing the f-vectors, homology groups, and relative homology groups of an arbitrary pair of simplicial complexes $\Gamma \subseteq \Delta$, see [68].) We have mentioned in Section 7 that the h-vector of a Cohen–Macaulay relative complex Δ/Γ is nonnegative. Equivalently, $h(\Delta) \geq h(\Gamma)$ (i.e., $h_i(\Delta) \geq h_i(\Gamma)$ for all i). Thus in particular by Corollary 7.3, we see that $h(\Delta) \geq h(\Gamma)$ if Δ and Γ are Cohen–Macaulay of the same dimension. This inequality was actually first proved by G. Kalai (unpublished) as part of his theory of algebraic shifting, and later by R. Adin [2, Thm. 6.5] by the same method used to prove Theorem 9.1 below. We can ask what happens when $\dim \Gamma < \dim \Delta$. There no longer needs to hold the inequality $h(\Delta) \geq h(\Gamma)$. For instance, if Δ is a one-dimensional simplex and Γ consists of the two vertices of Δ, then $h(\Delta) = (1, 0, 0, \ldots)$ and $h(\Gamma) = (1, 1, 0, \ldots)$. There is, however, a simple condition which assures that $h(\Delta) \geq h(\Gamma)$, whose proof provides a nice illustration of the technique of choosing a "special" l.s.o.p. This result first appeared in [169, Thm. 2.1].

9.1 Theorem. *Let Δ' be a Cohen–Macaulay subcomplex of the Cohen–Macaulay simplicial complex Δ. Let $e - 1 = \dim \Delta' \leq \dim \Delta = d - 1$. Assume that no set of $e + 1$ vertices of Δ' forms a face of Δ. (This last condition automatically holds if $d = e$.) Then $h_i(\Delta') \leq h_i(\Delta)$ for all i.*

Proof. If $F \in \Delta$, then let x^F denote the face monomial as in (7). Let I be the ideal of $k[\Delta]$ generated by all monomials x^F, where $F \notin \Delta'$. Clearly $k[\Delta'] \cong k[\Delta]/I$. It follows easily from Lemma 2.4 that when k is infinite we can find an l.s.o.p. $\theta_1, \ldots, \theta_d$ of $k[\Delta]$ with the following property: Each of $\theta_{e+1}, \ldots, \theta_d$ is a linear combination of vertices not contained in Δ'. In other words, the images of $\theta_{e+1}, \ldots, \theta_d$ in $k[\Delta']$ (when identified with $k[\Delta]/I$) are all zero. Let us call such an l.s.o.p. *special*.

Identify the special l.s.o.p. $\theta = \{\theta_1, \ldots, \theta_d\} \subset k[\Delta]_1$ with its image in $k[\Delta']$ (or, alternatively, think of $k[\Delta']$ as a $k[\Delta]$ module). Since $\theta_1, \ldots, \theta_d$ is special, it follows that $\theta_1, \ldots, \theta_e$ is an l.s.o.p. for $k[\Delta']$, and that

$$k[\Delta']/(\theta_1, \ldots, \theta_d) = k[\Delta']/(\theta_1, \ldots, \theta_e).$$

Now we have a degree-preserving surjection

$$R := k[\Delta]/(\theta) \xrightarrow{f} R' := k[\Delta']/(\theta) \cong R/I.$$

Since Δ and Δ' are Cohen–Macaulay we have $\dim_k R_i = h_i(\Delta)$ and $\dim_k R'_i = h_i(\Delta')$. The surjection $f : R_i \to R'_i$ shows that $h_i(\Delta) \geq h_i(\Delta')$, as desired. $\qquad\square$

9.2 Corollary. *Let Δ be a Cohen–Macaulay complex and $F \in \Delta$. Then $h_i(\mathrm{lk}(F)) \leq h_i(\Delta)$ for all i, where $\mathrm{lk}(F)$ denotes the link of F in Δ.*

Proof. Let

$$\mathrm{star}(F) = \{G \in \Delta : F \cup G \in \Delta\},$$

the *star* of F in Δ. Since Δ is Cohen–Macaulay, so is $\mathrm{lk}(F)$ by Reisner's theorem (Corollary II.4.2). Note that $\mathrm{star}(F)$ is just a cone over $\mathrm{lk}(F)$. Hence $\mathrm{star}(F)$ is also Cohen–Macaulay, and $h(\mathrm{star}(F)) = h(\mathrm{lk}(F))$. But $\mathrm{star}(F)$ satisfies the hypothesis of Theorem 9.1 (since $\dim(\mathrm{star}(F)) = \dim \Delta$ when Δ is Cohen–Macaulay or even just pure), and the proof follows. $\qquad\square$

For some additional work related to the results of this section, see [169].

10 Subdivisions

Intuitively, a subdivision (or triangulation) of a simplicial complex is obtained by dividing its faces up into smaller simplices. We will be concerned in this section with the connection between the f-vector of a simplicial complex Δ and of a subdivision of Δ. The results obtained depend on the precise meaning we give to the term "subdivision," so first we discuss several kinds of subdivisions.

In topology, one usually only deals with subdivisions of geometric realizations of simplicial complexes, and not of abstract simplicial complexes (e.g., [130, p. 83], [145, p. 121]). A *geometric simplex* is the convex hull of an affinely independent set of points in a (finite-dimensional) real vector space (or affine space). Suppose that Ω is a (finite) collection of geometric simplices in a real vector space such that (a) if $\sigma \in \Omega$ and τ is a face of σ, then $\tau \in \Omega$, and (b) if $\sigma, \tau \in \Omega$, then $\sigma \cap \tau$ is a face of both σ and τ. Thus the union X of all simplices in Ω is the geometric realization of a simplicial complex Δ whose faces correspond to the simplices in Ω. Let us call Ω a *geometric simplicial complex*, and Δ the *corresponding* abstract simplicial complex. A *subdivision* of Ω is a geometric simplicial complex Ω' such that every simplex in Ω' is contained in some simplex in Ω. Let Δ' be the abstract simplicial complex corresponding to Ω'. We say that Δ' is a *geometric subdivision* of Δ. (More generally, one can speak of *polyhedral subdivisions*, whose cells are arbitrary convex polytopes and not just simplices. In that case we should call a geometric subdivision into simplices a *simplicial subdivision* or *triangulation*. However, since we will be dealing only with simplicial complexes we will in general omit the adjective "simplicial" from the term "simplicial geometric subdivision.")

The condition that Δ' is a geometric subdivision of Δ is not easy to check, since one must construct the geometric simplicial complexes Ω and Ω' to which Δ and Δ' correspond. There is a more combinatorial notion of subdivision that seems to capture all the necessary information about geometric subdivisions related to f-vectors. First we define the more general notion of a "topological subdivision." (For an even more general type of subdivision, see [167, Def. 7.4].)

10.1 Definition. Let Δ be a simplicial complex. A pair (Δ', σ) is called a *topological subdivision* of Δ if Δ' is a simplicial complex and $\sigma : \Delta' \to \Delta$ satisfies the conditions:

(i) For every $F \in \Delta$, $\sigma^{-1}(2^F)$ is a subcomplex of Δ' whose geometric realization $|\sigma^{-1}(2^F)|$ is homeomorphic to a ball of dimension equal to $\dim F$ $(= \#F - 1)$.

(ii) $\sigma^{-1}(F)$ consists of the interior faces of the ball $\sigma^{-1}(2^F)$.

Often by abuse of notation we simply call Δ' a *subdivision* of Δ, the map σ being understood. For $F \in \Delta$ we call the subcomplex $\Delta'_F := \sigma^{-1}(2^F)$

the *restriction* of Δ' to F (or to the simplex 2^F). If $G \in \Delta'_F$ then we say that G *lies on* F. The unique smallest face $\sigma(G)$ of Δ on which G lies is called the *carrier* of G. Finally define the topological subdivision Δ' of Δ to be *quasi-geometric* if for every face $G \in \Delta'$, there does not exist a face $F \in \Delta$ for which (a) $\dim F < \dim G$, and (b) every vertex v of G lies on some subset (depending on v) of F. Equivalently, no face τ of Δ' has all its vertices on the closure 2^F of a face F of Δ of lower dimension than that of τ.

Clearly every geometric subdivision is quasi-geometric (since the vertices of a geometric simplex are affinely independent and therefore cannot belong to a geometric simplex of lower dimension). An example of a topological subdivision that is not quasi-geometric is given as follows: Let Δ be the simplex with vertices $1, 2, 3$. Add a vertex 4 inside the face 12 of Δ, and let the facets of Δ' be 123 and 124. (Imagine a curved edge inside Δ between vertices 1 and 2.) This subdivision is not quasi-geometric because the triangle (two-dimensional face) 124 of Δ' has all its vertices on the edge (one-dimensional face) 12 of Δ. For some examples of quasi-geometric subdivisions that are not geometric, see [167, p. 814].

Now let V be a d-element vertex set, and let 2^V denote the simplex with vertex set V. For any simplicial complex Δ, let $h(\Delta, x)$ denote its h-polynomial, i.e.,

$$h(\Delta, x) = \sum_{i=0}^{d} h_i(\Delta)x^i.$$

We come to the fundamental definition of this section.

10.2 Definition. For any topological subdivision (Γ, σ) of the simplex 2^V, define a polynomial $\ell_V(\Gamma, x) = \ell_0 + \ell_1 x + \cdots + \ell_d x^d$ by

$$\ell_V(\Gamma, x) = \sum_{W \subseteq V} (-1)^{\#(V-W)} h(\Gamma_W, x). \tag{31}$$

The polynomial $\ell_V(\Gamma, x)$ is called the *local h-polynomial* of Γ (or more accurately, of the pair (Γ, σ)), and the vector $(\ell_0, \ell_1, \ldots, \ell_d)$ is called the *local h-vector* of Γ.

As an example of the computation of a local h-vector, Let $\#V = 4$, and let $\Gamma = \mathrm{sd}(2^V)$, the first barycentric subdivision of the tetrahedron 2^V. Then $h(\Gamma_\emptyset) = 1$, $h(\Gamma_W) = 1$ if $\#W = 1$, $h(\Gamma_W) = 1 + x$ if $\#W = 2$, $h(\Gamma_W) = 1 + 4x + x^2$ if $\#W = 3$, and $h(\Gamma) = 1 + 11x + 11x^2 + x^3$. Hence

$$\begin{aligned}
\ell_V(\Gamma, x) &= 1 + 11x + 11x^2 + x^3 - 4(1 + 4x + x^2) + 6(1 + x) - 4 + 1 \\
&= x + 7x^2 + x^3,
\end{aligned}$$

so the local h-vector of Γ is given by $(0,1,7,1,0)$. For more information on $\mathrm{sd}(2^V)$ for arbitrary V, see [167, Prop. 2.4 and Prop. 4.20].

For a number of elementary properties and further examples of local h-vectors, see [167, §2]. We will present here the four main properties of local h-vectors. The first property explains the use of the term "local h-vector."

10.3 Theorem. (locality) *Let Δ be a pure $(d-1)$-dimensional simplicial complex, and let Δ' be a topological subdivision of Δ. Then*

$$h(\Delta', x) = \sum_{F \in \Delta} \ell_F(\Delta'_F, x) h(\mathrm{lk}_\Delta F). \tag{32}$$

Equation (32) shows that the contribution to $h(\Delta', x)$ "at the face F of Δ" is obtained from the behavior of Δ at F (the factor $h(\mathrm{lk}_\Delta F)$, which is independent of Δ') and the local h-vector of the subdivision of the simplex 2^F obtained by restricting Δ' to 2^F. The proof is a purely formal computation, as given in [167, Thm. 3.2].

10.4 Theorem. (reciprocity) *Let Γ be a topological subdivision of the simplex 2^V, where $\#V = d$. Then*

$$x^d \ell_V(\Gamma, 1/x) = \ell_V(\Gamma, x).$$

Equivalently, $\ell_i = \ell_{d-i}$ for all i.

Theorem 10.4 is a formal consequence of the fact that if Δ is a triangulation of a $(d-1)$-dimensional ball with h-vector $(h_0, h_1, \ldots, h_d)$ and if $\Delta/\partial\Delta$ is the relative simplicial complex consisting of the interior of Δ, then the h-vector of $\Delta/\partial\Delta$ is given by

$$h(\Delta/\partial\Delta) = (h_d, h_{d-1}, \ldots, h_0).$$

This equation is in turn an immediate consequence of Corollary II.7.2. For further details (in a much more general context), see [167, Thm. 3.3 and Cor. 7.7].

10.5 Theorem. (positivity) *Let Γ be a quasi-geometric subdivision of the simplex 2^V. Then $\ell_V(\Gamma) \geq 0$, i.e., $\ell_i \geq 0$ for all i.*

Proof (sketch). Theorem 10.5 is the most interesting property of local h-vectors from the standpoint of commutative algebra. The basic idea behind the proof is the following. Equation (31) expresses the polynomial $\ell_V(\Gamma, x)$ as an alternating sum, each term of which (ignoring the sign) is the Hilbert series of a graded $k[\Gamma]$-module $M_W = k[\Gamma_W]$. This suggests the existence of an exact sequence

$$k[\Gamma] \xrightarrow{\delta^0} \coprod_{\substack{W \subseteq V \\ \#W = d-1}} M_W \xrightarrow{\delta^1} \coprod_{\substack{W \subseteq V \\ \#W = d-2}} M_W \xrightarrow{\delta^2} \cdots \xrightarrow{\delta^{d-1}} M_\emptyset \to 0,$$

from which it will follow that the module $L = \ker(\delta^0)$ has a Hilbert function given by $H(L, i) = \ell_i \geq 0$. We now give some more details of this argument.

Given any topological subdivision Γ of 2^V, where $V = \{x_1, \ldots, x_d\}$, define an l.s.o.p. $\theta_1, \ldots, \theta_d$ of $k[\Gamma]$ to be *special* if each θ_i is a linear combination of vertices of Γ which do not lie on the face $V - \{x_i\}$ or its boundary. It is a simple consequence of Lemma 2.4(a) (see [167, Cor. 4.4] for details) that a special l.s.o.p. exists if and only if Γ is quasi-geometric. This is the crucial algebraic property of quasi-geometric subdivisions.

Now let $I_{\mathrm{int}(\Gamma)}$ denote the ideal of $k[\Gamma]$ generated by the face monomials x^F of interior faces F of Γ. (By Theorem II.7.3, $I_{\mathrm{int}(\Gamma)}$ is the canonical module of $k[\Gamma]$, but we don't need this fact here.) Suppose now that Γ is quasi-geometric and $\theta_1, \ldots, \theta_d$ is a special l.s.o.p. for $k[\Gamma]$. Then we denote the image of $I_{\mathrm{int}(\Gamma)}$ in $R_\Gamma := k[\Gamma]/(\theta_1, \ldots, \theta_d)$ by $L_V[\Gamma]$ and call it the *local face module* of Γ (with respect to V and $\theta_1, \ldots, \theta_d$). The ideal $L_V(\Gamma)$ is generated by homogeneous elements of R_Γ and thus has a grading

$$L_V(\Gamma) = L_0 \oplus L_1 \oplus \cdots \oplus L_d.$$

We claim that the Hilbert function of $L_V(\Gamma)$ is given by

$$\dim L_i = \ell_i,$$

from which it follows that $\ell_i \geq 0$.

Let N_i be the ideal of $k[\Gamma]$ generated by all monomials x^F for which $F \in \Gamma$ does not lie on the face $V - \{x_i\}$ of 2^V or on its boundary. Thus if $S = \{x_{i_1}, \ldots, x_{i_r}\} \subseteq V$ and $T = V - S$, then

$$k[\Gamma]/(N_{i_1} + \cdots + N_{i_r}) \cong k[\Gamma_T], \tag{33}$$

where Γ_T denotes the restriction of Γ to T. Define a chain complex $\mathcal{K}$ of $k[\Gamma]$-modules by

$$k[\Gamma] \xrightarrow{\varepsilon^0} \coprod_i k[\Gamma]/N_i \xrightarrow{\varepsilon^1} \coprod_{i<j} k[\Gamma]/(N_i + N_j) \xrightarrow{\varepsilon^2} \cdots \xrightarrow{\varepsilon^{d-1}} k[\Gamma]/(N_1 + \cdots + N_d) \to 0,$$

$$\tag{34}$$

where the map ε^r is defined as follows. If $u \in k[\Gamma]/(N_{i_1} + \cdots + N_{i_r})$, then

$$\varepsilon^r(u) = (u, -u, u, -u, \ldots) \in \prod_{j=1}^{d-r} k[\Gamma]/(N_{i_1} + \cdots + N_{i_r} + N_{a_j}),$$

where $\{a_1, \ldots, a_{d-r}\} = \{1, \ldots, d\} - \{i_1, \ldots, i_r\}$ and $a_1 < \cdots < a_{d-r}$. The complex $\mathcal{K}$ is easily seen to be exact [167, first paragraph of proof of Lemma 4.9]. Let $\theta = \{\theta_1, \ldots, \theta_d\}$ be a special l.s.o.p. for $k[\Gamma]$. We let $\mathcal{K}/\theta\mathcal{K}$ denote the complex obtained from $\mathcal{K}$ by modding out by $\theta_1, \ldots, \theta_d$, or equivalently, by tensoring $\mathcal{K}$ with $k[\Gamma]/\theta k[\Gamma]$. One then shows by a homological argument (due independently to K. Watanabe, S. Yuzvinsky, and M. Hochster, and given in [167, Lemma 4.9]) that exactness is preserved, i.e., the complex $\mathcal{K}/\theta\mathcal{K}$ is exact.

Let L' denote the kernel of the map $k[\Gamma]/\theta k[\Gamma] \rightarrow \coprod_i k[\Gamma]/(\theta k[\Gamma] + N_i)$ in $\mathcal{K}/\theta\mathcal{K}$. Thus L' is just the image of $N_1 \cap \cdots \cap N_d$ in $k[\Gamma]/\theta k[\Gamma]$. But $N_1 \cap \cdots \cap N_d = I_{\mathrm{int}(\Gamma)}$, so $L' = L_V(\Gamma)$. Since $\mathcal{K}/\theta\mathcal{K}$ is exact, it follows that the Hilbert series $F(L_V(\Gamma), x)$ of $L_V(\Gamma)$ is just the alternating sum of the Hilbert series of the modules appearing in $\mathcal{K}/\theta\mathcal{K}$. Hence

$$F(L_V(\Gamma), x) = \sum_{W \subseteq V} (-1)^{\#(V-W)} F(k[\Gamma_W]/\theta k[\Gamma_W], x).$$

But $k[\Gamma_W]$ is Cohen–Macaulay (since $|\Gamma_W|$ is a ball), and because θ is special it is an l.s.o.p. for $k[\Gamma_W]$ except for irrelevant elements θ_i which annihilate $k[\Gamma_W]$. Therefore

$$F(L_V[\Gamma], x) = h(\Gamma_W, x),$$

the h-polynomial of the simplicial complex Γ_W. Comparing with the definition (31) of $\ell_V(\Gamma, x)$ shows that $F(L_V[\Gamma], x) = \ell_V(\Gamma_W, x)$, as claimed.
$\square$

NOTE. The module $L_V(\Gamma)$ is an interesting object in its own right. For instance, it can be shown that $L_V(\Gamma)$ is "self-dual" in the sense that it is isomorphic to its canonical module $\Omega(L_V(\Gamma))$, as defined in Section I.12. This is the algebraic manifestation of Theorem 10.4 (the reciprocity theorem for local h-vectors).

NOTE. It is essential to assume in Theorem 10.5 that Γ is quasi-geometric, and not merely topological. For instance, let Γ be the subdivision of the simplex with vertex set $V = \{1, 2, 3, 4\}$ obtained by adding a vertex 5 inside face 123, and letting the facets of Γ be 1234 and 1235. Then $\ell_V(\Gamma, x) = -x^2$. This example is due to Clara Chan.

We are now in a position to give our main combinatorial application of local h-vectors. Recall from Section II.8 that a pure simplicial complex Δ is *Buchsbaum* if $\mathrm{lk}_\Delta F$ is Cohen–Macaulay for all $\emptyset \neq F \in \Delta$. In particular, Cohen–Macaulay complexes are Buchsbaum.

10.6 Theorem. *Let Δ' be a quasi-geometric subdivision of the Buchsbaum simplicial complex Δ. Then $h(\Delta') \geq h(\Delta)$.*

Proof. For a polynomial $p(x)$ with real coefficients, write $p(x) \geq 0$ to mean that every coefficient is nonnegative. Since $h(\mathrm{lk}_\Delta \emptyset) = \Delta$ and $\ell_\emptyset(\emptyset, x) = 1$, the term indexed by $F = \emptyset$ in (32) is just $h(\Delta, x)$. Since $\mathrm{lk}_\Delta F$ is Cohen–Macaulay for all $\emptyset \neq F \in \Delta$, we have for $F \neq \emptyset$ that $h(\mathrm{lk}_\Delta F, x) \geq 0$ (by Theorem II.3.3). On the other hand, by Theorem 10.5 we have $\ell_F(\Delta'_F, x) \geq 0$ for all $F \in \Delta$. Hence every term indexed by $F \neq \emptyset$ in the right-hand side of (32) is nonnegative, and the proof follows. $\square$

NOTE. The only subdivisions of a single vertex are trivial, so $\ell_F(\Delta'_F, x) = 0$ when $\#F = 1$. It is then apparent from the above proof that the condition that Δ is Buchsbaum in Theorem 10.6 can be relaxed to the condition that $\mathrm{lk}_\Delta F$ is Cohen–Macaulay for $\#F \geq 2$. For further information, see [167, Thm. 4.10].

NOTE. It is not known whether Theorem 10.6 continues to hold if we only assume that Δ' is a topological subdivision of Δ, as was conjectured (unpublished but mentioned in [167, Conj. 4.11]) by Kalai and this writer.

For our final basic property of local h-vectors, we need to define a special kind of geometric subdivision Γ of the simplex 2^V. We say that Γ is *regular* if it can be geometrically realized as a subdivision of a geometric simplex $\mathcal{S}$ with the following property: There exists a function $\omega : \mathcal{S} \to \mathbb{R}$ which is (a) piecewise-linear, i.e., ω is linear on each face of the subdivision, and (b) strictly convex, i.e., ω is convex $(\omega(\lambda u + (1 - \lambda)v) \geq \lambda \omega(u) + (1 - \lambda)\omega(v)$ for all $u, v \in \mathcal{S}$ and $0 \leq \lambda \leq 1)$ and is a *different* linear function on each facet (maximal face) of the subdivision. Note that regular subdivisions are by definition geometric and hence also quasi-geometric. There are numerous examples (e.g, [188, §5.1]) of geometric subdivisions that are not regular. For further information on the important concept of a regular subdivision (in the more general context of polyhedral subdivisions) see e.g. [19], [114], [188, §5.1].

10.7 Theorem. (unimodality) *Let Γ be a regular subdivision of the simplex 2^V, where $\#V = d$. Then the local h-vector $\ell_V(\Gamma) = (\ell_0, \dots, \ell_d)$ is unimodal, i.e. (since $\ell_i = \ell_{d-i}$),*

$$\ell_0 \leq \ell_1 \leq \cdots \leq \ell_{\lfloor d/2 \rfloor}.$$

Theorem 10.7 is a very deep result whose proof depends on the hard Lefschetz theorem for the decomposition theorem of the intersection homology of toric varieties. See [167, Thm. 5.2] for further details. It would be desirable to find a more elementary proof, perhaps using the machinery of McMullen's polytope algebra (which we mentioned at the end of Section 1 was used to prove Theorem 1.1 avoiding the use of toric varieties). It is unknown whether Theorem 10.7 continues to hold for geometric or even just quasi-geometric subdivisions.

There is an application of Theorem 10.7 which is analogous to the use of Theorem 10.5 to prove Theorem 10.6. We state only the most interesting case of this result here; a more general result is given by [167, Cor. 5.3]. The proof is completely analogous to that of Theorem 10.6 and will be omitted.

10.8 Theorem. *Let Δ be the boundary complex of a $(d-1)$-dimensional simplicial convex polytope, with h-vector $(h_0, h_1, \ldots, h_d)$. Let Δ' be a subdivision of Δ whose restriction to every face of Δ is regular. Let $h(\Delta') = (h'_0, h'_1, \ldots, h'_d)$. Then*

$$h_i - h_{i-1} \leq h'_i - h'_{i-1}, \quad 0 \leq i \leq \lfloor d/2 \rfloor.$$

It is natural to ask whether there are additional conditions on local h-vectors besides those given by Theorems 10.4, 10.5, and 10.7. For some results in this direction, see [167, Cor. 4.8], [50], [51].

There is a conjecture of Kalai and Stanley related to subdivisions that would imply that h-vectors of partitionable complexes (as defined in Section 2) increase under topological subdivision. Let Γ be a topological subdivision of the simplex 2^V, where $\#V = d$. Choose $1 \leq i \leq d$, and let Δ be a subcomplex of 2^V whose facets consist of i codimension one faces of 2^V. Let Γ_Δ denote the restriction of Γ to the subcomplex Δ. Then it is conjectured that the relative complex Γ/Γ_Δ satisfies

$$h_i(\Gamma/\Gamma_\Delta) > 0.$$

(We always have $h_j(\Gamma/\Gamma_\Delta) \geq 0$ for any j since the relative complex Γ/Γ_Δ is Cohen–Macaulay.)

Problems on Simplicial Complexes and their Face Rings

In the problems below, Δ always denotes a finite simplicial complex of dimension $d - 1$ with vertex set V and face ring $k[\Delta]$. Properties like "acyclic," "Cohen-Macaulay," etc., are always taken with respect to a fixed ground field k.

1. Let Δ be a simplicial complex with 159 3-dimensional faces. For $i \leq 2$, find the smallest possible number of i-faces of Δ. For $i \geq 4$, find the largest possible number of i-faces of Δ.

2. A $(d - 1)$-dimensional simplicial complex Δ is *pure* if all its facets (= maximal faces) have dimension $d - 1$. Let $(f_0, f_1, \ldots, f_{d-1})$ be the f-vector of a pure $(d - 1)$-dimensional simplicial complex.

   (a) Show that the vector $(f_{d-2}, f_{d-3}, \ldots, f_0, 1)$ is the f-vector of a simplicial complex.

   (b) Show that $f_i \leq f_{d-2-i}$ for $-1 \leq i \leq \lfloor (d - 3)/2 \rfloor$, and that $f_0 \leq f_1 \leq \cdots \leq f_{\lfloor (d-1)/2 \rfloor}$.

   (c) (unsolved) Characterize (or at least obtain significant new conditions on) the f-vector of a pure simplicial complex. (A complete characterization is probably hopeless, as it would give, for instance, all orders of finite projective planes.)

3. Give an example of a pure simplicial complex Δ which fails for some i to satisfy the "Upper Bound Inequality" $h_i \leq \binom{n-d+i-1}{i}$. Here $n = f_0(\Delta)$, the number of vertices of Δ. (There exists such an example with $h(\Delta) = (1, 2, 3, 4, 5, -26, 23, -8, 1)$.)

4. Give an example of two simplicial complexes Δ_1 and Δ_2 such that $|\Delta_1| \approx |\Delta_2|$, but such that $h(\Delta_1) \geq 0$ and $h(\Delta_2) \not\geq 0$. What is the smallest possible dimension of Δ_1 and Δ_2? (The symbol $\approx$ denotes homeomorphism of topological spaces.)

5. Let Δ be a $(d-1)$-dimensional simplicial complex. For $0 \leq j \leq d-1$, define the *j-skeleton* Δ_j of Δ by $\Delta_j = \{F \in \Delta : \dim F \leq j\}$. Express the h-vector $h(\Delta_{d-2})$ in terms of $h(\Delta)$.

6. Given integers $n > d \geq 1$, let $\Delta(n, d)$ be the simplicial complex with vertex set $[n]$ whose facets consist of all d-subsets F of $[n]$ with the

the following property: If $\{i, i+1, \ldots, j\} \subseteq F$, $i-1 \notin F$, $j+1 \notin F$, $i > 1$, and $j < n$, then $j - i$ is odd. Find the h-vector $h(\Delta(n, d))$ of $\Delta(n, d)$.

7. Let Γ and Δ be simplicial complexes on disjoint vertex sets V and W, respectively. Define the *join* $\Gamma * \Delta$ to be the simplicial complex on the vertex set $V \cup W$ with faces $F \cup G$, where $F \in \Gamma$ and $G \in \Delta$. (If Γ consists of a single point, then $\Gamma * \Delta$ is the *cone* over Δ. If Γ consists of two disjoint points, then $\Gamma * \Delta$ is the *suspension* of Δ.)

   (a) Compute the h-vector $h(\Gamma * \Delta)$ in terms of $h(\Gamma)$ and $h(\Delta)$.

   (b) The *boundary* $\Delta(d)$ *of the* $(d-1)$-*dimensional cross-polytope* (as an abstract simplicial complex) has vertex set $V = \{x_1, \ldots, x_d, y_1, \ldots, y_d\}$, with $F \subseteq V$ a face of $\Delta(d)$ if and only if $\{x_i, y_i\} \not\subseteq F$ for all i. Find $h(\Delta(d))$.

8. Let Δ be a triangulation of a $(d-1)$-sphere, $d > 3$.

   (a) Show that the link of every vertex is simply-connected (i.e., the fundamental group of $|\text{lk}(v)|$ is trivial for every vertex v).

   (b) (very difficult) Show that for $d \geq 6$, the link of an edge need not be simply-connected.

9. Let Δ be a $(d-1)$-dimensional simplicial complex.

   (a) Let $\sigma : \Delta \to \Delta$ be an automorphism of Δ of prime order p. Suppose that for all nonempty faces F of Δ we have $\sigma(F) \neq F$. Let $(h_0, \ldots, h_d)$ be the h-vector of Δ. Show that

   $$h_i(\Delta) \equiv (-1)^i \binom{d}{i} \pmod{p}.$$

   Deduce that $\tilde{\chi}(\Delta) \equiv -1 \pmod{p}$.

   (b) More generally, suppose G is a group of automorphisms of Δ, and let $\#G = g$. Suppose that for all $1 \neq \sigma \in G$ and $\emptyset \neq F \in \Delta$ we have $\sigma(F) \neq F$. Show that the congruences in (a) are still valid, with p replaced by g.

10. Let Δ be a $(d-1)$-dimensional simplicial complex with f-vector $(f_0, \ldots, f_{d-1})$ and h-vector $(h_0, \ldots, h_d)$. We say that Δ satisfies the

Dehn-Sommerville equations if $h_i = h_{d-i}$ for all i. Show that if d is even and Δ satisfies the Dehn-Sommerville equations, then

$$f_0 - f_1 + f_3 - f_4 + f_6 - f_7 + f_9 - f_{10} + \cdots = 0.$$

11. Show that there does not exist a triangulation Δ of a ball $\mathbb{B}^{d-1}$ for $d \geq 3$ such that every facet intersects the boundary in a $(d-2)$-face together with the opposite vertex.

12. Let Δ be a triangulation of a $(d-1)$-dimensional ball, with $h(\Delta) = (h_0, h_1, \ldots, h_d)$.

   (a) Show that

   $$\begin{aligned}
   h(\partial\Delta) = (\,&h_0 - h_d, h_0 + h_1 - h_d - h_{d-1}, \\
   &h_0 + h_1 + h_2 - h_d - h_{d-1} - h_{d-2}, \\
   &\ldots, h_0 + \cdots + h_{d-1} - h_d - \cdots - h_1).
   \end{aligned}$$

   (b) (unsolved) Show that $h_i \geq h_{d-i}$ for $0 \leq i \leq \lfloor d/2 \rfloor$.

   (c) Show that $h_i = h_{d-i}$ for $1 \leq i \leq d-1$ if and only if the boundary of Δ is the boundary of a simplex (i.e., has exactly d facets).

   (d) (unsolved) Characterize h-vectors of balls.

13. Let Δ be a pure $(d-1)$-dimensional simplicial complex. A face of Δ of dimension $< d-1$ is called *free* if it lies on exactly one facet.

   (a) If Δ is acyclic, must there exist a free face? What if Δ is contractible?

   (b) What if also Δ is shellable?

14. Show that if Δ is a shellable simplicial complex, then so is the barycentric subdivision $\mathrm{sd}(\Delta)$. (We can define $\mathrm{sd}(\Delta)$ to be the order complex of the poset of *nonempty* faces of Δ.)

15. Let Δ be a pure $(d-1)$-dimensional simplicial complex, with h-vector $h(\Delta) = (h_0, h_1, \ldots, h_d)$. Let Γ be the order complex of the poset of *nonempty* faces of Δ. Thus Γ is a balanced complex with colors $1, 2, \ldots, d$, where an i-dimensional face of Δ (regarded as a

vertex of Γ) is colored $i+1$. Let $S \subseteq [d]$. Show that the flag h-vector of Γ is given by

$$\beta_\Gamma(S) = \sum_{i=0}^{d} h_i D_{d+1}(\bar{S}, i+1),$$

where $D_{d+1}(\bar{S}, i+1)$ denotes the number of permutations $w = w_1 w_2 \cdots w_{d+1}$ of $[d+1]$ such that $\bar{S} := [d+1] - S = \{j : w_j > w_{j+1}\}$ (the *descent set* of w) and $w_{d+1} = i+1$.

16. Find explicitly every simplicial complex Δ with the property that *every* ordering of its facets is a shelling.

17. Let P be a finite graded poset with ranks $1, 2, \ldots, n$. Let $S \subseteq [n]$. Show that if P is shellable (i.e., the order complex $\mathcal{O}(P)$ is shellable), then so is the rank-selected subposet P_S.

18. A *d-pyramid* $\mathcal{P}$ is the convex hull of a $(d-1)$-polytope $\mathcal{Q}$, called the *basis* of $\mathcal{P}$, and a point $x \notin$ aff $\mathcal{Q}$ (= the affine span of $\mathcal{Q}$), called the *apex* of $\mathcal{P}$. Show that

$$f_i(\mathcal{P}) = f_i(\mathcal{Q}) + f_{i-1}(\mathcal{Q}),$$

with the convention $f_{-1}(\mathcal{Q}) = f_{d-1}(\mathcal{Q}) = 1$.

19. Let $\mathcal{Q}$ be a $(d-1)$-polytope and I a closed line segment such that $\mathrm{relint}(\mathcal{Q}) \cap \mathrm{relint}(I)$ is a single point. (Here relint denotes relative interior.) Then the d-polytope $\mathcal{P} = \mathrm{conv}(\mathcal{Q} \cup I)$ is called a *d-bipyramid* with *basis* $\mathcal{Q}$. Show that

$$\begin{aligned} f_i(\mathcal{P}) &= f_i(\mathcal{Q}) + 2f_{i-1}(\mathcal{Q}), \quad 0 \le i \le d-2 \\ f_{d-1}(\mathcal{P}) &= 2f_{d-2}(\mathcal{Q}). \end{aligned}$$

20. Show that the Euler relation $f_0 - f_1 + \cdots + (-1)^{d-1} f_{d-1} = 1 + (-1)^{d-1}$ is the most general linear relation holding among the components of the f-vector of a d-polytope. Equivalently, the affine span in $\mathbb{R}^d$ of all f-vectors of d-polytopes has codimension 1 (or dimension $d-1$). HINT: Use the previous two problems.

21. Show that the Dehn-Sommerville equations are the most general linear relations holding among the components of the f-vector (or h-vector) of a *simplicial d-polytope*.

22. A convex d-polytope $\mathcal{P}$ is *cubical* if every facet (or equivalently every proper face) is combinatorially equivalent to a cube of the appropriate dimension. Let $\mathcal{C}_d$ denote the set of all f-vectors of cubical d-polytopes.

   (a) Show that the affine span aff $\mathcal{C}_d$ of $\mathcal{C}$ has dimension at most $\lfloor d/2 \rfloor$.

   (b) By constructing suitable examples, show that in fact aff $\mathcal{C}_d$ has dimension exactly $\lfloor d/2 \rfloor$.

   (c) Show that a cubical polytope of dimension at least four has an even number of vertices. (The conclusion is false for three-dimensional cubical polytopes.)

   (d) (surprisingly difficult) Show that a cubical d-polytope has at least 2^d vertices.

   (e) Find a cubical 4-polytope for which $f_3 > f_0$.

   (f) (unsolved) Given d and n, what is the maximum number of i-dimensional faces of a cubical d-polytope with n vertices?

23. (generalizing (d) above) Suppose that the d-polytope $\mathcal{P}$ has no triangular 2-face. Show that for every facet F of $\mathcal{P}$, there is a facet G disjoint from F. Deduce that $\mathcal{P}$ has at least 2^d vertices.

24. Here we give two unsolved problems due to Kalai which illustrate our abysmal ignorance of certain aspects of convex polytopes.

   (a) Show that if $\mathcal{P}$ is a centrally-symmetric d-polytope, then $f_{-1} + f_0 + \cdots + f_{d-1} \geq 3^d$.

   (b) Let j be a positive integer. Show that for d sufficiently large, every d-polytope has a j-face which is either a simplex or (is combinatorially equivalent to) a cube. (Kalai has shown that this statement is true for $j = 2$.)

25. Let Δ be an acyclic (over a field k, say) $(d-1)$-dimensional simplicial complex. Let

$$F(x) = \sum_{i=-1}^{d-1} f_i(\Delta) x^{i+1}.$$

   (a) Show that $F(x)$ is divisible by $x + 1$.

(b) Let $F(x) = (x + 1)G(x)$. Show that $G(x)$ has nonnegative coefficients. (HINT: Let $\cdots \rightarrow C_1 \rightarrow C_0 \rightarrow C_{-1} \rightarrow 0$ be the augmented chain complex used to define the reduced homology groups $\tilde{H}_i(\Delta; k)$. What is the rank of the boundary operator $\partial_i : C_i \rightarrow C_{i-1}$?)

(c) Show that Δ can be partitioned into pairwise disjoint 2-element sets $\{F, F'\}$ such that $F \subset F'$ and $|F' - F| = 1$. (Note that this implies (b).) (HINT: Let G be the Hasse diagram of the face poset $P(\Delta)$ of Δ, regarded as an undirected graph. G is a bipartite graph with vertex bipartition (V_0, V_1), where

$$V_i = \{F \in \Delta \ : \ |F| \equiv i \pmod 2\}.$$

We want to find a complete matching M in G. By the Marriage Theorem, M will exist if (and only if) every subset S of V_0 is connected to at least $|S|$ vertices in V_1. Assume $S \subseteq V_0$ fails to satisfy this condition. Deduce that the k-vector space kS with basis S contains a cycle (in the sense of reduced homology) which fails to be a boundary.)

(d) Show that one can choose the partitioning into sets $\{F, F'\}$ in (c) so that the faces F form a subcomplex of Δ. Deduce that $(f_0, \ldots, f_{d-1})$ is the f-vector of some acyclic simplicial complex Δ if and only if the polynomial $G(x) = 1 + g_0 x + \cdots + g_{d-2} x^{d-1}$ of (b) has the property that $(g_0, \ldots, g_{d-2})$ is the f-vector of some simplicial complex.

26. (a) Generalize the previous exercise as follows. A *Betti set* of a simplicial complex Δ is a subset $B \subseteq \Delta$ such that for all i,

$$\{F \in B : \dim F = i\} = \tilde{\beta}_i(\Delta),$$

where $\tilde{\beta}_i(\Delta) = \dim \tilde{H}_i(\Delta; k)$, the ith reduced Betti number of Δ (over k). Show that Δ can be written as a disjoint union $\Delta = \Delta' \cup B \cup \Omega$, where (i) Δ' is a subcomplex of Δ, (ii) B is a Betti set, (iii) B is an *antichain*, i.e., if $F, G \in B$ and $F \subseteq G$, then $F = G$, (iv) $\Delta' \cup B$ is a subcomplex of Δ, and (v) there exists a bijection $\eta : \Delta' \rightarrow \Omega$ such that for all $F \in \Delta'$ we have $F \subset \eta(F)$ and $|\eta(F) - F| = 1$.

(b) Deduce a characterization of the f-vector of a simplicial complex with given Betti numbers, generalizing Exercise 25(d).

27. (a) A simplicial complex Δ is called *j-acyclic* if for every face F of Δ with $|F| \leq j$, the link of F is acyclic. (Thus 0-acyclic just means that Δ is acyclic.) Let $\mathbf{f}(\Delta) = (f_0, f_1, \ldots, f_{d-1})$. Show that if Δ is *j-acyclic*, then the polynomial $D_\Delta(x) := 1 + \sum_0^{d-1} f_i(\Delta)x^{i+1}$ is divisible by $(1 + x)^{j+1}$.

    (b) Show that $\mathbf{f} = (f_0, f_1, \ldots, f_{d-1})$ is the f-vector of a *j-acyclic* simplicial complex if and only if there is a simplicial complex Δ' such that $\mathbf{f} = \mathbf{f}(\sigma * \Delta)$, where σ is a *j*-dimensional simplex and $*$ denotes join. Equivalently, we have

    $$D_\Delta(x) = (1 + x)^{j+1} \left(1 + \sum_0^{d-j-2} g_i x^{i+1} \right),$$

    where $(g_0, g_1, \ldots, g_{d-j-2})$ is the f-vector of some simplicial complex.

    (c) (unsolved) Can every *j-acyclic* simplicial complex Δ (regarded as a poset ordered by inclusion) be partitioned into disjoint intervals $[F, F']$ of length (or rank) $j + 1$? (This would imply that $D_\Delta(x)/(1 + x)^{j+1}$ is a polynomial with nonnegative coefficients.) Moreover, can the intervals $[F, F']$ be chosen so that the set of their bottom elements F form a subcomplex of Δ? (This would imply (b).)

28. A convex polytope $\mathcal{P}$ is *j-neighborly* if every j vertices of $\mathcal{P}$ lie on a proper face of $\mathcal{P}$. Let $\mathcal{P}$ be a *j*-neighborly polytope with at least $j + 1$ vertices. Show that every j vertices of $\mathcal{P}$ are affinely independent. Deduce that $\mathcal{P}$ is *i*-neighborly for all $1 \leq i \leq j$.

29. Let Δ be a *j*-neighborly Euler complex of dimension $d - 1$, with $j > \lfloor d/2 \rfloor$. Show that Δ is the boundary of a simplex.

30. Show that every *j*-neighborly $2j$-polytope is simplicial.

31. Suppose that Δ and Δ' are simplicial complexes whose face rings $k[\Delta]$ and $k[\Delta']$ are isomorphic as graded algebras (or even as rings). Show that Δ and Δ' are isomorphic.

32. A graded algebra $k[x_1, \ldots, x_n]/I$ is said to be a *complete intersection* if the ideal I is generated by a regular sequence. Find all simplicial complexes Δ for which the face ring $k[\Delta]$ is a complete intersection.

33. Give a direct proof that depth $k[\Delta] = 1$ if Δ is a disconnected simplicial complex.

34. Show that the depth of $k[\Delta]$ is equal to the largest value of j for which the $(j-1)$-skeleton Δ_{j-1} is Cohen-Macaulay.

35. Show that if Δ_1 and Δ_2 are Cohen-Macaulay simplicial complexes, then so is their join $\Delta_1 * \Delta_2$.

36. (a) Given $1 \le d \le n$, let $\Delta_{n,d}$ be the simplicial complex whose vertex set V consists of the squares of an $n \times d$ chessboard (so $|V| = nd$), and whose faces consist of subsets F of V with no two elements in the same row or column. Compute the f-vector and h-vector of $\Delta_{n,d}$.

   (b) Show that $\Delta_{n,d}$ is Cohen-Macaulay if and only if $2d \le n + 1$.

37. Give an example of a (finite) simplicial complex Δ and a field k for which $k[\Delta]$ has no l.s.o.p.

38. Let $\Delta = [G_1, F_1] \cup \cdots \cup [G_r, F_r]$ be a partitioning of the Cohen-Macaulay simplicial complex Δ. Let $\theta_1, \ldots, \theta_d$ be an l.s.o.p. of $k[\Delta]$. Show that the monomials x^{G_i} need not form a basis for $k[\Delta]$ as a (free) module over $k[\theta_1, \ldots, \theta_d]$. (There exists an example with $d = 2$.)

39. Using Reisner's theorem, show that a Cohen-Macaulay complex is pure.

40. (a) Let P be a Cohen-Macaulay poset (i.e., the order complex $\Delta(P)$ is Cohen-Macaulay) of dimension $d - 1$, and let $\hat{P} = P \cup \{\hat{0}, \hat{1}\}$. Let μ denote the Möbius function of $\hat{P}$. Write $\mathrm{type}(P)$ as short for $\mathrm{type}(\Delta(P))$. Suppose that for every $x < y$ in $\hat{P}$, either the order complex $\Delta(x, y)$ of the open interval (x, y) is doubly Cohen-Macaulay or else $\mu(x, y) = 0$. Show that

$$\mathrm{type}(P) = (-1)^{d-1} \sum \mu(x_0, x_1)\mu(x_1, x_2) \cdots \mu(x_{r-1}, x_r),$$

where the sum is over all chains $\hat{0} = x_0 < x_1 < \cdots < x_r = \hat{1}$ in $\hat{P}$ such that $\mu(x_{i-1}, x_{i+1}) = 0$ for all i.

   (b) Show that (finite) modular lattices satisfy the conditions of part (a).

(c) Suppose P is a distributive (or even meet-distributive) lattice. A chain $\hat{0} = x_0 < x_1 < \cdots < x_r = \hat{1}$ in P is called a *Loewy chain* if every interval $[x_{i-1}, x_i]$ is a boolean algebra. Show that type(P) is equal to the number of minimal Loewy chains in P (i.e., the number of chains C (containing $\hat{0}$ and $\hat{1}$) which are Loewy chains, but for which any proper subset of C is not a Loewy chain).

(d) Let $t(a, b) = \text{type}(C_{a+1} \times C_{b+1})$, where C_m denotes an m-element chain. Let

$$F(x, y) = \sum_{a,b \geq 0} t(a, b) x^a y^b.$$

Show that

$$F(x, y) = \frac{1 - xy}{1 - x - y + x^2 y^2}.$$

Bibliography

[1] W. W. Adams and P. Loustaunau, *An Introduction to Gröbner Bases*, Graduate Studies in Math. Vol. III, American Mathematical Society, Providence, RI, 1994.

[2] R. Adin, Combinatorial structure of simplicial complexes with symmetry, Ph.D. thesis, Hebrew University, Jerusalem, 1991.

[3] R. Adin, On face numbers of rational simplicial polytopes with symmetry, *Advances in Math.* **115** (1995), 269–285.

[4] R. Adin, On h-vectors and symmetry, *Contemporary Math.* **178** (1994), 1–20.

[5] H. Anand, V. C. Dumir, and H. Gupta, A combinatorial distribution problem, *Duke Math. J.* **33** (1966), 757–769.

[6] M. F. Atiyah and I.G. Macdonald, *Introduction to Commutative Algebra*, Addison-Wesley, Reading, Massachusetts, 1969.

[7] K. Baclawski, Cohen-Macaulay ordered sets, *J. Algebra* **63** (1980), 226–258.

[8] K. Baclawski, Rings with lexicographic straightening law, *Advances in Math.* **39** (1981), 185–213.

[9] K. Baclawski, Cohen–Macaulay connectivity and geometric lattices, *European J. Combinatorics* **3** (1982), 293–305.

[10] K. Baclawski, Canonical modules of partially ordered sets, *J. Algebra* **83** (1983), 1–5.

[11] K. Baclawski and A. M. Garsia, Combinatorial decompositions of a class of rings, *Advances in Math.* **39** (1981), 155–184.

[12] M. M. Bayer and L. J. Billera, Generalized Dehn-Sommerville relations for polytopes, spheres and Eulerian partially ordered sets, *Invent. math.* **79** (1985), 143–157.

[13] I. Bárány and L. Lovász, Borsuk's theorem and the number of facets of centrally symmetric polytopes, *Acta Math. Acad. Sci. Hung.* **40** (1982), 323–329.

[14] T. Becker, V. Weispfennig, and H. Kredel, *Gröbner Bases: A Computational Approach to Commutative Algebra*, Graduate Texts in Mathematics, vol. 141, Springer-Verlag, New York/Berlin/-Heidelberg, 1993.

[15] D. H. Bernstein and A. Iarrobino, A nonunimodal graded Gorenstein Artin algebra in codimension five, *Comm. Algebra* **20** (1992), 2323–2336.

[16] L. J. Billera, Polyhedral theory and commutative algebra, in *Mathematical Programming: The State of the Art* (A. Bachem, M. Grötschel, and B. Korte, eds.), Springer-Verlag, New York/Berlin, 1983, pp. 57–77.

[17] L. J. Billera, Homology of smooth splines: Generic triangulations and a conjecture of Strang, *Trans. Amer. Math. Soc.* **310** (1988), 325–340.

[18] L. J. Billera, The algebra of continuous piecewise polynomials, *Advances in Math.* **76** (1989), 170–183.

[19] L. J. Billera, P. Filliman, and B. Sturmfels, Constructions and complexity of secondary polytopes, *Advances in Math.* **83** (1990), 155–179.

[20] L. J. Billera and C. W. Lee, Sufficiency of McMullen's conditions for f-vectors of simplicial polytopes, *Bull. Amer. Math. Soc.* **2** (1980), 181–185.

[21] L. J. Billera and C. W. Lee, A proof of the sufficiency of McMullen's conditions for f-vectors of simplicial polytopes, *J. Combinatorial Theory (A)* **31** (1981), 237–255.

[22] L. J. Billera and L. L. Rose, Gröbner basis methods for multivariate splines, in *Mathematical Methods in Computer Aided Geometric*

Design (T. Lyche and L. L. Schumaker, eds.), Academic Press, New York, 1989, pp. 93–104.

[23] L. J. Billera and L. L. Rose, A dimension series for multivariate splines, *Discrete Comput. Geom.* **6** (1991), 107–128.

[24] L. J. Billera and L. L. Rose, Modules of piecewise polynomials and their freeness, *Math. Z.* **209** (1992), 485–497.

[25] A. Björner, Some matroid inequalities, *Discrete Math.* **31** (1980), 101–103.

[26] A. Björner, Shellable and Cohen-Macaulay partially ordered sets, *Trans. Amer. Math. Soc.* **260** (1980), 159–183.

[27] A. Björner, Some combinatorial and algebraic properties of Coxeter complexes and Tits buildings, *Advances in Math.* **52** (1984), 173–212.

[28] A. Björner, Coxeter groups and combinatorics, in *Proc. XIXth Nordic Congress of Mathematicians, Reykyavik, 1984* (J. Stefánsson, ed.), Icelandic Math. Soc., Reykjavik, 1985, pp. 24–32.

[29] A. Björner, Posets, regular CW complexes and Bruhat order, *European J. Combinatorics* **5** (1984), 7–16.

[30] A. Björner, The homology and shellability of matroids and geometric lattices, in [186], pp. 226–283.

[31] A. Björner, Topological methods, in *Handbook of Combinatorics* (R. Graham, M. Grötschel, and L. Lovász, eds.), North-Holland, Amsterdam, to appear.

[32] A. Björner and K. Eriksson, Extendable shellability for rank 3 matroid complexes, *Discrete Math.* **132** (1994), 373–376.

[33] A. Björner, P. Frankl, and R. Stanley, The number of faces of balanced Cohen-Macaulay complexes and a generalized Macaulay theorem, *Combinatorica* **7** (1987), 23–34.

[34] A. Björner, A. Garsia, and R. Stanley, An introduction to the theory of Cohen-Macaulay posets, in *Ordered Sets* (I. Rival, ed.), Reidel, Dordrecht/Boston/London, 1982, pp. 583–615.

[35] A. Björner and G. Kalai, An extended Euler-Poincaré theorem, *Acta Math.* **161** (1988), 279–303.

[36] A. Björner and G. Kalai, Extended Euler-Poincaré relations for cell complexes, in *Applied Geometry and Discrete Mathematics: The Victor Klee Festschrift*, DIMACS Series in Discrete Mathematics and Theoretical Computer Science (P. Gritzmann and B. Sturmfels, eds.), vol. 4, 1991, pp. 81–89.

[37] A. Björner and M. Wachs, Bruhat order of Coxeter groups and shellability, *Advances in Math.* **43** (1982), 87–100.

[38] A. Björner and M. Wachs, On lexicographically shellable posets, *Trans. Amer. Math. Soc.* **277** (1983), 323–341.

[39] A. Björner and M. Wachs, Shellable nonpure complexes and posets. *Trans. Amer. Math. Soc.*, to appear.

[40] M. Boij, *Artin Level Algebras*, Doctoral Dissertation, Royal Institute of Technology, Stockholm, 1994.

[41] M. Boij, Graded Gorenstein Artin algebras whose Hilbert functions have a large number of valleys, *Comm. Algebra* **23** (1995), 97–103.

[42] M. Boij and D. Laksov, Nonunimodality of graded Gorenstein Artin algebras, *Proc. Amer. Math. Soc.* **120** (1994), 1083–1092.

[43] N. Bourbaki, *Groupes et algèbres de Lie*, Chap. 4, 5, et 6, Hermann, Paris, 1968.

[44] A. Brøndsted, *An Introduction to Convex Polytopes*, Graduate Texts in Mathematics **90**, Springer-Verlag, New York/Berlin, 1983.

[45] K. S. Brown, *Buildings*, Springer-Verlag, New York/Berlin, 1989.

[46] J. I. Brown and C. J. Colbourn, Roots of the reliability polynomial, *SIAM J. Discrete Math.* **5** (1992), 571–585.

[47] M. Bruggesser and P. Mani, Shellable decompositions of cells and spheres, *Math. Scand.* **29** (1971), 197–205.

[48] W. Bruns and J. Herzog, *Cohen-Macaulay Rings*, Cambridge Studies in Advanced Mathematics **39**, Cambridge University Press, Cambridge, 1993.

[49] B. Buchberger, Gröbner bases — an algorithmic method in polynomial ideal theory, Chapter 6 in *Multidimensional Systems Theory* (N. K. Bose, ed.), Reidel, Dordrecht/Boston, 1985.

[50] C. S. Chan, *On shellings and subdivisions of convex polytopes*, Ph.D. thesis, M.I.T., 1992.

[51] C. S. Chan, On subdivisions of simplicial complexes: Characterizing local *h*-vectors, *Discrete Comput. Geom.* **11** (1994), 465–476.

[52] M. K. Chari, Two decompositions in topological combinatorics with applications to matroid complexes, preprint dated February 7, 1995.

[53] M. K. Chari, Matroid inequalities, *Discrete Math.*, to appear.

[54] R. Charney, Metric geometry: connections with combinatorics, in *DIMACS Series in Discrete Mathematics and Theoretical Computer Science*, to appear.

[55] R. Charney and M. Davis, The Euler characteristic of a nonpositively curved, piecewise linear Euclidean manifold, preprint.

[56] R. Courant, Variational methods for the solution of problems of equilibrium and vibration, *Bull. Amer. Math. Soc.* **49** (1943), 1–23.

[57] G. Clements and B. Lindström, A generalization of a combinatorial theorem of Macaulay, *J. Combinaorial Theory* **7** (1969), 230–238.

[58] D. Cox, J. Little, and D. O'Shea, *Ideals, Varieties and Algorithms*, Undergraduate Texts in Mathematics, Springer, New York, 1992.

[59] G. Danaraj and V. Klee, A representation of 2-dimensional pseudomanifolds and its use in the design of a linear-time shelling algorithm, *Ann. Discrete Math.* **2** (1978), 53–63.

[60] V. I. Danilov, The geometry of toric varieties, *Russian Math. Surveys* **33** (1978), 97–154. Translated from *Uspekhi Mat. Nauk.* **33** (1978), 85–134.

[61] C. De Concini, D. Eisenbud, and C. Procesi, Young diagrams and determinantal varieties, *Inventiones Math.* **56** (1980), 129–165.

[62] C. De Concini, D. Eisenbud, and C. Procesi, Hodge algebras, *Astérisque* **91** (1982), 1–87,

[63] P. Diaconis and A. Gangolli, Rectangular arrays with fixed margins, in *Discrete Probability and Algorthims* (D. Aldous, *et al.*, eds.), IMA vol. 72, Springer-Verlag, Berlin/Heidelberg/New York, 1995, pp. 15–41.

[64] P. Diaconis and B. Sturmfels, Algebraic algorithms for sampling from conditional distributions, *Ann. Stat.*, to appear.

[65] T. A. Dowling, On the independent set numbers of a finite matroid, *Ann. Discrete Math.* **8** (1980), 21–28.

[66] A. M. Duval, *Simplicial posets: f-vectors and free resolutions*, Ph.D. thesis, M.I.T., 1991.

[67] A. M. Duval, A combinatorial decomposition of simplicial complexes, *Israel J. Math.* **87** (1994), 77–87.

[68] A. M. Duval, On f-vectors and relative homology, preprint.

[69] E. Ehrhart, Sur un problème de géometrie diophantienne linéaire, I, II, *Crelle J. (= J. Reine Angew. Math.)* **226** (1967), 1–29; **227** (1967), 25–49.

[70] E. Ehrhart, Démonstration de la loi de réciprocité du polyèdre rationnel, *C.R. Acad. Sci. Paris* **265A** (1967), 91–94.

[71] E. Ehrhart, Sur les carrés magiques, C.R. Acad. Sci. Paris **277A** (1973), 575–577.

[72] D. Eisenbud, Introduction to algebras with straightening laws, in *Ring Theory and Algebra III* (B. McDonald, ed.), Lecture Notes in Pure and Applied Mathematics **55**, Dekker, New York, 1980, pp. 243–268.

[73] G. Ewald, *Combinatorial Convexity and Algebraic Geometry*, Graduate Texts in Mathematics, Springer-Verlag, New York/Berlin, in preparation.

[74] J. Fine, The Mayer-Vietoris and *IC* equations for convex polytopes, *Discrete Comput. Geom.* **13** (1995), 177–188.

[75] J. Folkman, The homology groups of a lattice, *J. Math. Mech.* **15** (1966), 631–636.

[76] P. Frankl, Z. Füredi, and G. Kalai, Shadows of colored complexes, *Math. Scand.* **63** (1988), 169–178.

[77] W. Fulton, *An Introduction to Toric Varieties*, Annals of Mathematics Studies, no. 131, Princeton University Press, Princeton, NJ, 1993.

[78] W. Fulton, *Young Tableaux with Applications to Representation Theory and Combinatorics*, Cambridge University Press, Cambridge, 1995.

[79] W. Fulton and J. Harris, *Representation Theory*, Graduate Texts in Mathematics **129**, Springer-Verlag, New York/Berlin/Heidelberg, 1991.

[80] A. M. Garsia, Méthodes combinatoire dans la théorie des anneaux de Cohen-Macaulay, *C. R. Acad. Sci. Paris Sér. A* **288** (1979), 371–374.

[81] A. M. Garsia, Combinatorial methods in the theory of Cohen-Macaulay rings, *Advances in Math.* **38** (1980), 229–266.

[82] A. M. Garsia and D. Stanton, Group actions on Stanley-Reisner rings and invariants of permutation groups, *Advances in Math.* **51** (1984), 107–201.

[83] P. Goosens, Shelling of tilings, IMA preprint series #486, Institute for Mathematics and Its Applications, Minneapolis, 1989.

[84] H.-G. Gräbe, The canonical module of a Stanley-Reisner ring, *J. Algebra* **86** (1984), 272–281.

[85] C. Greene and D.J. Kleitman, Proof techniques in the theory of finite sets, in *Studies in Combinatorics* (G.-C. Rota, ed.), Mathematical Association of America, 1978, pp. 22–79.

[86] B. Grünbaum, *Convex Polytopes*, Wiley, London-New York-Sydney, 1967.

[87] R. Hartshorne, *Local Cohomology*, Lecture Notes in Math., no. 41, Springer, Berlin-Heidelberg-New York, 1967.

[88] J. Herzog and E. Kunz (eds.), *Der kanonische Modul eines Cohen–Macaulay-Rings*, Lecture Notes in Math., no. 238, Springer, Berlin-Heidelberg-New York, 1971.

[89] T. Hibi, Level rings and algebras with straightening laws, *J. Algebra* **117** (1988), 343–362.

[90] T. Hibi, Face number inequalities for matroid complexes and Cohen-Macaulay types of Stanley-Reisner rings of distributive lattices, *Pacific J. Math.* **154** (1992), 253–264.

[91] T. Hibi, *Algebraic Combinatorics on Convex Polytopes*, Carslaw Publications, Glebe, Australia, 1992.

[92] P. J. Hilton and U. Stammbach, *A Course in Homological Algebra*, Springer-Verlag, Berlin-Heidelberg-New York, 1971.

[93] P. J. Hilton and Y.-C. Wu, *A Course in Modern Algebra*, Wiley, New York, 1974.

[94] M. Hochster, Rings of invariants of tori, Cohen–Macaulay rings generated by monomials, and polytopes, *Annals of Math.* **96** (1972), 318–337.

[95] M. Hochster, Cohen–Macaulay rings, combinatorics, and simplicial complexes, in *Ring Theory II* (Proc. Second Oklahoma Conference) (B.R. McDonald and R. Morris, ed.), Dekker, New York, 1977, pp. 171–223.

[96] M. Hochster and J. L. Roberts, Rings of invariants of reductive groups acting on regular rings are Cohen–Macaulay, *Advances in Math.* **13** (1974), 115–175.

[97] W. V. D. Hodge, Some enumerative results in the theory of forms, *Proc. Camb. Phil. Soc.* **39** (1943), 22–30.

[98] W. V. D. Hodge and D. Pedoe, *Methods of Algebraic Geometry*, vol. 2, Cambridge University Press, Cambridge, 1968.

[99] J. E. Humphreys, *Reflection Groups and Coxeter Groups*, Cambridge Studies in Advanced Mathematics **29**, Cambridge University Press, Cambridge, 1990.

[100] A. Iarrobino, Vector spaces of forms I: Ancestor ideals, preprint.

[101] A. A. Iarrobino, Associated graded algebra of a Gorenstein Artin algebra, *Memoirs Amer. Math. Soc.*, vol. 107, no. 514, American Mathematical Society, Providence, RI, 1994.

[102] D. M. Jackson and G. H. J. van Rees, The enumeration of generalized double stochastic nonnegative integer square matrices, *SIAM J. Comput.* **4** (1975), 474–477.

[103] R.-Q. Jia, The conjecture of Stanley for symmetric magic squares, in *Formal Power Series and Algebraic Combinatorics*, Proceedings of the Fifth Conference (Florence, June 21–25, 1993) (A. Barlotti, M. Delest, and R. Pinzani, eds.), University of Florence, pp. 193–300.

[104] R.-Q. Jia, Symmetric magic squares and multivariate splines, preprint.

[105] G. Kalai, *f*-vectors of acyclic complexes, *Discrete Math.* **55** (1984), 97–99.

[106] G. Kalai, Many triangulated spheres, *Discrete Comput. Geom.* **3** (1988), 1–14.

[107] I. Kaplansky, *Commutative Rings*, revised ed., University of Chicago Press, Chicago/London, 1974.

[108] B. Kind and P. Kleinschmidt, Schälbare Cohen-Macaulay-Komplexe und ihre Parametrisierung, *Math. Z.* **167** (1979), 173–179.

[109] S. L. Kleiman and D. Laksov, Schubert calculus, *Amer. Math. Monthly* **79** (1972), 1061–1082.

[110] P. Kleinschmidt, Über Hilbert-Funktionen graduierter Gorenstein-Algebren, *Arch. Math.* **43** (1984), 501–506.

[111] P. Kleinschmidt and Z. Smilansky, New results for simplicial spherical polytopes, in *Discrete and Computational Geometry* (J. E. Goodman, R. Pollack, and W. Steiger, eds.), DIMACS Series in Discrete Mathematics and Theoretical Computer Science, vol. 6, 1991, pp. 187–197.

[112] F. Knop, Der kanonische Modul eines Invariantenringes, *J. Algebra* **127** (1989), 40–54.

[113] F. S. Macaulay, Some properties of enumeration in the theory of modular systems, *Proc. London Math. Soc.* **26** (1927), 531–555.

[114] C. Lee, Regular triangulations of convex polytopes, in *Applied Geometry and Discrete Mathematics: The Victor Klee Festschrift*, DIMACS Series in Discrete Mathematics and Theoretical Computer Science (P. Gritzmann and B. Sturmfels, eds.), vol. 4, 1991, pp. 443–456.

[115] W. B. R. Lickorish, Unshellable triangulations of spheres, *European J. Combinatorics* **12** (1991), 527–530.

[116] L. Lovász, *Combinatorial Problems and Exercises*, North-Holland, Amsterdam/New York/ Oxford, 1979.

[117] I. G. Macdonald, The volume of a lattice pohyhedron, *Proc. Cambridge Philos. Soc.* **59** (1963), 719–726.

[118] I. G. Macdonald, Polynomials associated with finite cell complexes, *J. London Math. Soc.* (2) **4** (1971), 181–192.

[119] P. A. MacMahon, *Combinatory Analysis*, vols. 1–2, Cambridge, 1916; reprinted by Chelsea, New York, 1960.

[120] H. Matsumura, *Commutative Algebra*, second edition, Benjamin/ Cummings, Reading, MA, 1980.

[121] H. Matsumura, *Commutative Ring Theory*, Cambridge University Press, Cambridge, 1986.

[122] P. McMullen, The maximum numbers of faces of a convex polytope, *Mathematika* **17** (1970), 179–184.

[123] P. McMullen, The numbers of faces of simplicial polytopes, *Israel J. Math.* **9** (1971), 559–570.

[124] P. McMullen, Valuations and Euler-type relations for certain classes of convex polytopes, *Proc. London Math. Soc.* **35** (1977), 113–135.

[125] P. McMullen, On simple polytopes, *Invent. math.*, to appear.

[126] P. McMullen, Weights on polytopes, preprint.

[127] P. McMullen and G. C. Shephard, *Convex Polytopes and the Upper Bound Conjecture*, London Math. Soc. Lecture Note Series **3**, Cambridge University Press, Cambridge, 1971.

[128] M. Miyazaki, Characterization of Buchsbaum complexes, *Manuscripta Math.* **63** (1989), 245–254.

[129] T. S. Motzkin, Comontone curves and polyhedra, Abstract 111, *Bull. Amer. Math. Soc.* **63** (1957), 35.

[130] J. R. Munkres, *Elements of Algebraic Topology*, Addison-Wesley, Reading, MA, 1984.

[131] J. Munkres, Topological results in combinatorics, *Michigan Math. J.* **31** (1984), 113–128.

[132] G. Nicoletti and N. White, Axiom systems, in [184], Ch. 2, pp. 29–44.

[133] I. Novik, *Upper Bound Theorems for simplicial manifolds*, Master's thesis, Hebrew University, 1996.

[134] T. Oda, *Convex Bodies and Algebraic Geometry (An Introduction to the Theory of Toric Varieties)*, Springer-Verlag, Berlin/New York, 1988.

[135] C. Peskine, private communication, December, 1994.

[136] G. Pick, *Geometrisches zur Zahlenlehre, Naturwissenschaft Zeitschrift Lotos*, Prague, 1899.

[137] G. Pólya, Untersuchungen über Lüken und Singularitäten von Potenzreihen, *Math. Zeitschrift* **29** (1928–29), 549–640.

[138] T. Popoviciu, Asupra unei probleme de partitie a numerelor, *Studie şi cercetari ştiintifice, Akad. R.P.R. Filiala Cluj* **4** (1953), 7–58.

[139] G. Reisner, Cohen–Macaulay quotients of polynomial rings, *Advances in Math.* **21** (1976), 30–49.

[140] R. M. Robinson, Integer-valued entire functions, *Trans. Amer. Math. Soc.* **153** (1971), 451–468.

[141] M. E. Rudin, An unshellable triangulation of a tetrahedron, *Bull. Amer. Math. Soc.* **64** (1958), 90–91.

[142] P. Schenzel, On the number of faces of simplicial complexes and the purity of Frobenius, *Math. Z.* **178** (1981), 125–142.

[143] M.-P. Schützenberger, A characteristic property of certain polynomials of E. F. Moore and C. E. Shannon, in *RLE Quarterly Progress Report. No. 55*, Research Laboratory of Electronics, M.I.T., 1959, 117–118.

[144] R. Simon, *The combinatorial properties of "cleanness,"*, Ph.D. thesis, Univ. of Bielefeld, 1992.

[145] E. H. Spanier, *Algebraic Topology*, McGraw-Hill, New York, 1966.

[146] R. Stanley, Ordered structures and partitions, *Memoirs of the Amer. Math. Soc.*, no. 119 (1972), iii + 104 pages.

[147] R. Stanley, Linear homogeneous diophantine equations and magic labelings of graphs, *Duke Math. J.* **40** (1973), 607–632.

[148] R. Stanley, Combinatorial reciprocity theorems, *Advances in Math.* **14** (1974), 194–253.

[149] R. Stanley, Cohen–Macaulay rings and constructible polytopes, *Bull. Amer. Math. Soc.* **81** (1975), 133–135.

[150] R. Stanley, The Upper Bound Conjecture and Cohen–Macaulay rings, *Studies in Applied Math.* **54** (1975), 135–142.

[151] R. Stanley, Magic labelings of graphs, symmetric magic squares, systems of parameters, and Cohen–Macaulay rings, *Duke Math. J.* **43** (1976), 511–531.

[152] R. Stanley, Cohen–Macaulay complexes, in *Higher Combinatorics* (M. Aigner, ed.), Reidel, Dordrecht and Boston, 1977, pp. 51–62.

[153] R. Stanley, Some combinatorial aspects of the Schubert calculus, in *Combinatoire et Réprésentation du Groupe Symétrique (Strasbourg, 1976)*, Lecture Notes in Math., no. 579, Springer-Verlag, Berlin, 1977, pp. 217–251.

[154] R. Stanley, Generating functions, in *Studies in Combinatorics* (G.-C. Rota, ed.), Mathematical Assoc. of America, 1978, pp. 100–141.

[155] R. Stanley, Hilbert functions of graded algebras, *Advances in Math.* **28** (1978), 57–83.

[156] R. Stanley, Balanced Cohen-Macaulay complexes, *Trans. Amer. Math. Soc.* **249** (1979), 139–157.

[157] R. Stanley, Weyl groups, the hard Lefschetz theorem, and the Sperner property, *SIAM J. Algebraic and Discrete Methods* **1** (1980), 168–184.

[158] R. Stanley, Decompositions of rational convex polytopes, *Annals of Discrete Math.* **6** (1980), 333–342.

[159] R. Stanley, The number of faces of a simplicial convex polytope, *Adances in Math.* **35** (1980), 236–238.

[160] R. Stanley, Linear diophantine equations and local cohomology, *Inv. math.* **68** (1982), 175–193.

[161] R. Stanley, The number of faces of simplicial polytopes and spheres, in *Discrete Geometry and Convexity* (J. E. Goodman, et al., eds.), Ann. New York Acad. Sci., vol. 440 (1985), pp. 212–223.

[162] R. Stanley, *Enumerative Combinatorics*, vol. 1, Wadsworth and Brooks/Cole, Pacific Grove, CA, 1986; to be reprinted by Cambridge University Press.

[163] R. Stanley, On the number of faces of centrally-symmetric simplicial polytopes, *Graphs and Combinatorics* **3** (1987), 55–66.

[164] R. Stanley, Generalized *h*-vectors, intersection cohomology of toric varieties, and related results, in *Commutative Algebra and Combinatorics* (M. Nagata and H. Matsumura, eds.), Advanced Studies in Pure Mathematics **11**, Kinokuniya, Tokyo, and North-Holland, Amsterdam/New York, 1987, pp. 187–213.

[165] R. Stanley, *f*-vectors and *h*-vectors of simplicial posets, *J. Pure Applied Algebra* **71** (1991), 319–331.

[166] R. Stanley, On the Hilbert function of a graded Cohen-Macaulay domain, *J. Pure Applied Algebra* **73** (1991), 307–314.

[167] R. Stanley, Subdivisions and local *h*-vectors, *J. Amer. Math. Soc.* **5** (1992), 805–851.

[168] R. Stanley, A combinatorial decomposition of acyclic simplicial complexes, *Discrete Math.* **120** (1993), 175–182.

[169] R. Stanley, A monotonicity property of h-vectors and h^*-vectors, *European J. Combinatorics* **14** (1993), 251–258.

[170] R. Stanley, A survey of Eulerian posets, in *Polytopes: Abstract, Convex, and Computational* (T. Bisztriczky, P. McMullen, R. Schneider, and A. I. Weiss, eds.), NATO ASI Series C, vol. 440, Kluwer, Dordrecht/Boston/London, 1994, pp. 301–333.

[171] R. Stanley, Flag f-vectors and the cd-index, *Math. Z.* **216** (1994), 483–499.

[172] R. Stanley, Flag-symmetric and locally rank-symmetric partially ordered sets, *Electronic J. Combinatorics*, to appear.

[173] R. Stanley, Graph colorings and related symmetric functions: Ideas and applications, *Discrete Math.*, to appear.

[174] G. Strang, Piecewise polynomials and the finite element method, *Bull. Amer. Math. Soc.* **79** (1973), 1128–1137.

[175] G. Strang, The dimension of piecewise polynomial spaces and one-sided approximation, in *Proc. Conf. Numerical Solution of Differential Equations (Dundee, 1973)*, Lecture Notes in Mathematics, vol. 363, Springer-Verlag, New York/Heidelberg/Berlin, 1974, pp. 144–152.

[176] J. Stückrad and W. Vogel, *Buchsbaum Rings and Applications*, Springer-Verlag, Berlin/Heidelberg/New York, 1986.

[177] B. Sturmfels, *Gröbner Bases and Convex Polytopes*, American Mathematical Society, Providence, RI, to appear.

[178] B. Sturmfels and N. White, Stanley decompositions of the bracket ring, *Math. Scand.* **67** (1990), 183–189.

[179] J. Tits, *Buildings of Spherical Type and Finite BN-Pairs*, Lecture Notes in Mathematics, no. 386, Springer-Verlag, Berlin/Heidelberg/New York, 1974.

[180] G. X. Viennot, Heaps of pieces, I: Basic definitions and combinatorial lemmas, in *Combinatoire énumerative* (G. Labelle and P. Leroux, eds.), Lecture Notes in Mathematics, no. 1234, Springer-Verlag, Berlin/Heidelberg/New York, 1986, pp. 321–350.

[181] J. W. Walker, *Topology and combinatorics of ordered sets*, Ph.D. thesis, M.I.T., 1981.

[182] K. Watanabe, Certain invariant subrings are Gorenstein. I, *Osaka J. Math.* **11** (1974), 1–8.

[183] K. Watanabe, Certain invariant subrings are Gorenstein. I, *Osaka J. Math.* **11** (1974), 1–8.

[184] N. White, ed., *Theory of Matroids*, Encyclopedia of Mathematics and Its Applications, vol. 26, Cambridge University Press, Cambridge, 1986.

[185] N. White, ed., *Combinatorial Geometries*, Encyclopedia of Mathematics and Its Applications, vol. 29, Cambridge University Press, Cambridge, 1987.

[186] N. White, ed., *Matroid Applications*, Encyclopedia of Mathematics and Its Applications, vol. 40, Cambridge University Press, Cambridge, 1992.

[187] E. C. Zeeman, On the dunce hat, *Topology* **2** (1964), 341–358.

[188] G. M. Ziegler, *Lectures on Polytopes*, Springer-Verlag, New York/Berlin/Heidelberg, 1995.

Index

Progress in Mathematics

Edited by:

H. Bass
Dept. of Mathematics
Columbia University
New York, NY 10010
U.S.A.

J. Oesterlé
Dép. de Mathématiques
Université de Paris VI
4, Place Jussieu
75230 Paris Cedex 05
France

A. Weinstein
Dept. of Mathematics
University of California
Berkeley, CA 94720
U.S.A.

Progress in Mathematics is a series of books intended for professional mathematicians and scientists, encompassing all areas of pure mathematics. This distinguished series, which began in 1979, includes authored monographs and edited collections of papers on important research developments as well as expositions of particular subject areas.

We encourage preparation of manuscripts in some form of TeX for delivery in camera-ready copy which leads to rapid publication, or in electronic form for interfacing with laser printers or typesetters.

Proposals should be sent directly to the editors or to: Birkhäuser Boston, 675 Massachusetts Avenue, Cambridge, MA 02139, U. S. A.

Printed in the United States
70755LV00001B/17